Energy in Perspective

Mark Cronshaw

Energy in Perspective

 Springer

Mark Cronshaw
Principal, Resource Economics LLC
Boulder, CO, USA

EnergyInPerspective Supplement can be downloaded from https://www.springer.com/us/book/978-3-030-63540-4

ISBN 978-3-030-63543-5   ISBN 978-3-030-63541-1   (eBook)
https://doi.org/10.1007/978-3-030-63541-1

This Springer imprint is published by the registered company Springer Nature Switzerland AG
The registered company address is: Gewerbestrasse 11, 6330 Cham, Switzerland

"What all the world desires to have—power" [1]
22 March 1776, Matthew Boulton, who together with James Watt commercialized the steam engine.

---

[1] https://todayinsci.com/QuotationsCategories/S_Cat/SteamEngine-Quotations.htm

Preface

Winters are cold in Afghanistan. In 2007, as part of my work as an energy expert, I drove north from the capital, Kabul, over the mountainous Salang Pass to gas fields near the village of Sheberghan. Only 6% of Afghans had grid-supplied electricity at the time, and many did not have access to hot water. The gas had potential to provide heat and electricity. Some years later, I was in Dhaka, Bangladesh, working on regulatory enhancements to deal with a looming energy shortage. The power went out as I was making a presentation. The room went dark and the air-conditioning stopped working. It became hotter and hotter as we waited for the power to return.

It is easy to overlook the benefits of energy when it is available and reliable. But energy is not always reliable. About 55 million people in Canada and the northeastern USA lost power for about 7 hour in 2003 due to cascading technical failures. It made national news, but it was small compared to the 620 million people in northern and eastern India who were left without power in 2012 due to power blackouts.

Energy has consequences as well as benefits. You have doubtless read about them. The carbon dioxide concentration in the earth's atmosphere is about 400 parts per million, the highest it has been in the last 400,000 years. The concentration has increased rapidly since the industrial revolution due to the widespread use of fossil fuels for transportation, heating, cooking, electricity generation, and other industrial activities. The global temperature has increased, glaciers have shrunk, sea ice has been lost, sea-level rise is accelerating, ice on rivers and lakes is breaking up earlier in the spring, plant and animal habitat has changed, and severe weather seems more commonplace.

Like any industrial activity, there are risks associated with energy. In 2010, the Deepwater Horizon oil rig that was drilling a well for BP in the Gulf of Mexico suffered a blowout. A huge amount of oil spilled into the Gulf. Eleven workers died. Other large oil spills have occurred when oceangoing tankers ran aground near Alaska, France, and England.

Oil is not the only risky energy source. There have been several coal mine disasters over the years in China, Turkey, and the USA. Nuclear and hydroelectricity have the potential to provide clean energy, but even they are not without risk. Releases of radiation have occurred at Three Mile Island in the USA, Chernobyl in Ukraine, and the Fukushima Daiichi nuclear plant in Japan. Also, nuclear waste disposal presents a challenge because the waste will remain radioactive for thousands of years. Hydroelectric dams displace people from their homes prior to reservoir filling and they limit fish migration. And when hydroelectric dams fail, the loss of life can be catastrophic: about 171,000 people died in China in 1975 when the Banqiao Dam failed.[1]

Why do we tolerate such risks and environmental impacts? There is a simple answer. Energy is really useful. Imagine a winter without heat or light, or think about returning to an era with travel only by horse or wagon. It is tough to cover long distances that way! You are fortunate if you have reliable energy. About 1.1 billion people (1/7 of the world's population) do not have access to electricity, and another 2.8 billion rely on wood or other biomass for cooking and heating. It takes time and effort to gather firewood and other biomass, which limits the hours available for productive work. This tends to lead to a cycle of economic poverty. Furthermore, burning of wood and other biomass creates indoor and outdoor air pollution that causes about 4.3 million deaths per year.[2] Without access to electricity, people often rely on kerosene lamps for light. These lamps provide poor light, contribute to air pollution, and can cause fires. Houses and health centers without electricity lack refrigeration, hindering preservation of food, medicines, and vaccines.

I wrote this book to raise awareness of energy issues. Do you think about the energy that enables you to drive cars, take warm showers, dry your hair, cook, preserve your food in fridges/freezers, and heat/cool your homes? Where does it come from? How does energy use vary around the world? How have energy supplies and usage changed over time? How are they likely to change in the future? These are important topics that deserve careful analyses. This book presents science- and technology-based facts to establish a foundation for such analyses. It discusses how energy is produced, transformed, and consumed. Energy is used for many purposes including transportation, cooking, heating, cooling, lighting, electricity, petrochemicals, and other industry. It comes in many forms (oil, gas, coal, hydropower, solar, and wind) and must be delivered from its sources to its end users. The book provides insight at a level that is readily accessible to an inquisitive nonexpert and may also be of interest to experts given its breadth of coverage. It provides technical perspectives on each of the sectors. It also considers the ownership structure of firms that operate in these sectors. It reviews how energy sources have changed over time and makes an educated guess about how energy systems may change in the future.

You have inherited energy systems from the past. You can be a voice about their evolution. For example, coal-fired power plants emit a lot of carbon dioxide as an intrinsic consequence of generating electricity. One way to reduce their emissions is to stop using them. But what will make up for the lost electricity? Renewables? Nuclear energy? Energy efficiency? Natural gas?

Energy use has both benefits and hazards. We tolerate the risks because energy provides goods, services, and enormous convenience. Have we achieved the right balance between benefit and risk? In my opinion, it is not possible to answer that question or to develop rational energy policy without simultaneous consideration of

---

[1] https://en.wikipedia.org/wiki/Banqiao_Dam, retrieved Feb. 9, 2017.
[2] http://www.worldbank.org/en/topic/energy/overview#1, retrieved Feb. 8, 2017.

both benefits and risks. The goal of this book is to provide a firm foundation for thinking about it. The text and end-of-chapter exercises include several topics that show how data combined with critical thinking skills can provide insight.

Together, you and I can form a partnership that draws on my experiences with energy and relies on your actions, big and small, as you help to shape our collective future. Since you are reading this book, you have some interest in the topic. I am thrilled. Energy is really important. Please learn things from this book. Research topics that interest you. Take courses on subjects that the book skates over (for lack of space). Read material with different perspectives, even if they do not seem to agree with your own. Different perspectives are a gift: either you will adjust your beliefs to become better aligned with new facts or you will have more confidence about your knowledge when your openness to the possibility of improved understanding does not actually result in change. Form your own opinions and take action based on all that you know. Have discussions with your family and friends. Let us make decisions that are good for us, for our kids, for our planetary coinhabitants, and for the future.

Boulder, CO, USA   Mark Cronshaw

Acknowledgments

This book would not have been possible without support from many sources. The Internet was an enormous benefit for my research. I have used data from many sources including the BP Statistical Review of World Energy, the US Energy Information Agency, and the International Energy Agency. BP and those agencies have worked tirelessly for years to provide energy data. I trust that I have interpreted these data accurately, but if not, I am responsible for any misunderstanding.

I am grateful to my former employers for giving me the opportunity to work on many fascinating projects, and to my colleagues for their collaboration. In particular, I want to thank John Gustavson and Edwin Moritz for enabling me to return to the energy sector after a 20-year hiatus.

I learned many skills, both technical and analytical, from my teachers at universities in the UK and the USA. My parents emphasized the importance of learning and encouraged me all along the way. At my birth in Massachusetts, none of us would have imagined that my education would take me to boarding schools in the UK, to Cambridge, and then back to the USA for several postgraduate degrees. My now-grown children grew up in the home of a professional student. My heartfelt thanks to them.

Carron Meaney, my fiancée, deserves enormous gratitude for her patience. Not everyone is chastised for letting the electric kettle boil furiously, nor is everyone exposed to a differential equation for tracking the roasting progress of the Thanksgiving turkey. I am so lucky to have her in my life. She is remarkably gracious and a wonderful person to travel with.

This book is dedicated to my father, who had to support himself from the age of 16. He was a fantastic role model, manifesting great compassion, curiosity, and human kindness.

Contents

List of Figures

List of Tables

Introduction

1

1.1 Daily Energy

The alarm wakes you from your slumber. Maybe you are the sort of person who wakes up full of energy, refreshed after a night's sleep. Or perhaps you wake up slowly, grudgingly reaching to turn off the alarm. The air in your bedroom feels cold, but it is warm in bed. Without you even knowing it, your body has been releasing energy during the night. Thank goodness for the blankets that provide an insulating layer keeping heat inside the covers rather than letting it all escape to the room. Heat always moves from a hotter place to a cooler one. You can slow the rate of heat transfer, but its escape is inevitable.

Thank goodness also for the electricity that provides energy to the alarm clock, ensuring that the alarm goes off at the preset time. And even in a groggy state you might be grateful for the warmth in the house. The central heating system has been circulating hot air through the house, so that it is much warmer indoors than outdoors. Energy both heats the air and drives the fan that circulates it. Natural gas often does the heating, while electricity drives the fan. You could have chosen much quieter electric baseboard heating; it uses electricity to heat up resistance elements in the heaters and needs no fan to circulate the heat. Conveniently, the hot air from the baseboard heaters is less dense than the cooler air in the room, so it rises above the cooler air. This is natural convection, rather than forced convection from the fan. Either way, the heat inside the house escapes to the cooler outdoors. Perhaps you should invest in better insulation for your walls, windows, and ceiling to slow down the inexorable heat escape.

Lying in bed feels cozy, but it will not get the day's activities done. So you reach to turn on a light in preparation for getting up. You do not spare a moment's thought for the convenience of flicking a switch to turn on a light. But it is much easier than ensuring that a kerosene lantern was full and reaching for a match to light it, or maintaining an inventory of candles to provide the light. The candles can be used later in the evening to provide light as you set a romantic mood. This morning, the energy that powers the lights probably comes from electricity generated miles away

M. Cronshaw, *Energy in Perspective*,
https://doi.org/10.1007/978-3-030-63541-1_1

and delivered to your house by transmission and distribution lines, unless you have solar panels. Even then, you will probably draw electricity from the grid in the morning since the sun is not up yet.[1]

You walk to the bathroom, delighting in the warmth of the floor that has been heated by resistance heaters. First things first: a moment on the toilet. That done, the toilet flushes, also using energy as the contents are whisked away. That energy is cleverly stored in the form of potential energy in the toilet tank. The water in the tank is above the bowl, so gravity makes it fall from the tank into the bowl and down through the indoor plumbing. Unlike the water in the tank that does disappear, the potential energy in the water does not vanish. It is transformed into kinetic energy as the water flows. But how does the water in the tank replenish itself? Potential energy again serves its purpose, as water from storage facilities in your neighborhood is delivered to your house with sufficient pressure that it fills the tank when it is empty. So the neighborhood water storage facilities work just like your toilet tank but at a different scale.

Although your mind may still be working slowly as you are not yet fully awake, you might wonder how the water in the neighborhood storage tanks gets replenished. Energy is required to do that. Typically, pumps driven by electric motors will have been used to fill them. Electricity has been pretty useful in getting your day started. It is too early now to figure out where it came from, but there will be plenty to say about that later.

You turn on the shower, taking care to avoid the initial cold stream of water. Momentarily, the warm water appears. It is a rare person who enjoys a cold shower! It takes energy to warm the water. Water is usually heated by natural gas or electricity. It is common to have a storage tank full of hot water. It too loses its heat to its cooler surroundings, but only slowly if the tank is properly insulated. You might have a tank-less heater that heats water just as it is needed. Whether the water heater is tank-less or not, it takes some time for the cold water in the pipes supplying the shower to be replaced by warm water which makes the shower enticing.

After the shower, more personal hygiene remains. You may shave with an electric razor and/or dry your hair. You are using more energy: electricity for the electric devices, and gas or electricity for the warm water. You may use a curling iron, using electricity to generate heat.

As you get dressed you may think both about how you wish to appear during the day and about appropriate thermal insulation during the cool seasons (to manage that heat loss), or lack of insulation to stay cool during the hot seasons. Once dressed, it is time for breakfast. Energy is central to this activity also. Your coffee machine uses electricity to heat the water. You may use a kettle instead. If on the stove, it will heat the water using gas or electricity. Of course an electric kettle uses electricity. Your electric kettle may have clear sides. If so, you will see bubbles

---

[1]You might have battery backup attached to your solar system, just in case the grid power is not working. It is reassuring to be able to store energy at home, just in case the grid is down. Engineers are working diligently to figure out how to store more energy at reasonable cost in reasonably sized devices.

appear at the base of the liquid and throughout the water once the water has come to a rolling boil. You might notice a different sound as the water boils. You may think that a good cup of coffee or tea requires a vigorous rolling boil in the kettle. Surprise! The temperature of the water does not change while the water is boiling. It takes energy to heat the water, and it also takes energy for the water to change phase from liquid to vapor (steam). The temperature of the water rises while it is being heated, but it does not change when the two phases (liquid and vapor) coexist while boiling. The phase change requires energy and does release moisture (steam) to your room. You may want to decide if the moisture is worth the energy required in order to generate it. The energy costs money and causes more carbon emissions than would occur otherwise. This is because electricity is provided by generation facilities, of which there are many types including coal, gas, oil, hydroelectric, nuclear, wind, and solar. Coal, gas, and oil are fossil fuels. The chemical energy in these fuels is released by burning them with air. This burning process is known as combustion. The combustion transforms the fuels into water and carbon dioxide. Carbon dioxide concentrations in the atmosphere have increased to unprecedented levels and are associated with global warming.

The heat that results from the combustion in the generation facilities, far removed from your house, is used to produce electricity. The more electricity that is generated, the more fuel that must be burned. So the more steam you generate in your kettle, the more electricity you use and the more carbon that is emitted at generation facilities (unless your electricity is generated by renewables, nuclear, or hydro[2]). The extra steam has no effect on the temperature of the water in the kettle. One way for you to reduce your carbon footprint is not to let your kettle continue to boil! Your tea or coffee will taste just the same without the extra boiling.

You need more than tea or coffee for your breakfast. Any milk, eggs, or yoghurt that you have in your refrigerator needs energy to keep them cool. The toaster uses electrical energy to transform your bread or bagel into a more palatable form. Your egg (and perhaps bacon) requires energy to cook, either gas or electricity. The radio or TV that informs, amuses, or annoys you during breakfast requires energy. Your cell phone will have been charged overnight thanks to the electricity that was delivered. It stores enough energy (if you are lucky) to work throughout the day. Clever batteries!

Returning to the bathroom after breakfast to brush your teeth, you will use energy if you have an electric toothbrush. You may use a manual toothbrush. They work fine, but isn't it convenient to let the electric motor in the toothbrush do the scrubbing? Another clever battery stores enough energy in the electric toothbrush so that you can use it unencumbered by wires. But electricity delivers energy to the rechargeable batteries. You get the idea that there are many ways to do things, ranging from purely manual to energy-intensive methods. Energy provides convenience, particularly when it is provided in the form of electricity. There is no mess to clean up (unlike wood ash from a wood stove), although the out-of-sight electricity generation is certainly having an impact on the environment.

---

[2] See Chap. 7 for a discussion.

Now that you are fed, dressed, and cleaned up, you are ready to face the world. You may go to class, drive the kids to school, drive yourself to work, drive to do some errands, or drive to do something fun like a yoga class. Yes, there is a common theme: driving, another activity for which energy provides enormous convenience. No need to saddle up the horse or hitch the horses to a wagon. Instead you simply turn the key in the ignition. All cars store energy. All cars have a battery. Gasoline- or diesel-fueled cars use their battery only to turn the starter motor to initiate running of the car. After that initial start, the gasoline or diesel fuel provides all the energy needed by the car. Hybrid cars also use gasoline as their energy source, but they are careful about how they use the energy. For example, instead of wasting energy by turning it into heat in the brakes while slowing down, it uses the energy that is liberated during braking to charge a battery.

A combustion process also occurs in the car's engine. The chemical energy stored in the fuel is released as the fuel is converted into carbon dioxide and water. If it is cold, you may notice water leaking from the exhaust pipe of the car in front of you. You will not notice it later when the car is warm, since the water will be vaporized by the heat generated during combustion and will leave the tail pipe as invisible steam. You will never see the carbon dioxide that is coming out of the tail pipe since it is invisible.

There are many possible destinations for your car journey, perhaps school, work, gym, shop, or government office. All of these would be classified as commercial facilities, unless you go to an industrial location such as a factory. You will benefit from energy in any commercial facility that you visit. The facility will be heated or air-conditioned, ventilated, and lit. All of those use energy. Many will have kitchens, some with refrigerators. All of those also use energy. Many will have computers and other office equipment, all of which are powered with electricity. Even an unplugged laptop computer had its batteries charged up. You may go to a transportation facility like an airport, a train station, or a ferry building. Each of those uses energy for heating, air-conditioning, lighting, and ventilation. The planes, trains, or boats also use energy. It is much easier to sit in your seat (or even stand) than to expend your own energy to walk, run, or paddle to your destination! If you happen to park in a particularly cold location, you may plug your car into an electrical source during winter to power a small engine block heater so that it will be easy to start your vehicle when you return.

If your destination is an industrial facility you will witness energy consumption on an industrial scale (pun intended). The largest US industrial consumers of energy are refineries, chemical plants, paper mills, and iron/steel mills. This does not include fossil fuel-powered electrical generation facilities that also consume industrial quantities of energy. Well, to say "consume" is not quite right. Rather, they transform the chemical energy in the fossil fuel into electrical energy. Electrical energy is very convenient for users. There is no need to store it, and it is available at the flick of a switch. This is much easier than having to shovel coal into a boiler, load wood into a stove, or check your inventory of fuel oil or propane and order a top-up when necessary. Also refineries do not consume much energy. They receive crude oil and process it into refined products, mostly gasoline, diesel, and jet fuel that are used in transportation. They do use up some energy as part of the processing. Chemical plants also use some energy in their processing as they transform feedstocks that could be used directly for energy but have higher value as products such as ethane (a precursor for polyethylene, much used as plastic bags) or propane that can be conveniently stored as a liquid for use in residential heating, gas grills, or transportation, or further processed into polypropylene for use in packaging, furniture, or underwear. However, the bulk of the feedstock to chemical plants is not used as energy, but rather as precursors for the products.

When you return home from your destination, you will be glad to enter your warm (or air-conditioned) home, to be able to prepare a meal, and to be entertained by TV, music, or a book (impossible to enjoy without lighting). Energy provides enormous convenience for living. It enables us to be warm or cool, to move between locations, to prepare food, to store and preserve food at low temperatures, and to facilitate many activities with power tools, like lawn mowers. We tend to hear a great deal about carbon emissions and their impact on climate. We do not hear as much about the sources of those emissions. Societies tolerate such emissions because they are the by-product of energy use that provides us with substantial benefits. An intelligent discussion of environmental impacts must include a consideration of both the sources of the emissions and their benefits. There is an intrinsic trade-off between the benefits and the environmental impacts.

This book will increase your awareness of the use of energy. It describes how energy is used in our daily lives, how energy use has changed over time, and how it differs across countries in the world. Energy is used for heating, cooling, doing work, and transportation. It is generated from fossil fuels (coal, gas, and oil), hydroelectric dams, nuclear power plants, solar, and wind. The book discusses the relative amounts used in the residential, commercial,[3] transportation, and industrial sectors.

1.2   My Energy Bills

I live in the mountains of Colorado. My house was built in the 1980s in a location that receives a great deal of sun. I have installed energy-efficient windows that limit heat loss during the winter and heat gain in the summer. I do not have air-conditioning. Over a recent 12-month period I spent almost $1000 for gas and electricity as shown in Table 1.1.

The table shows that I used more than twice as much energy in the form of gas as in electricity (43,677 MJ compared to 20,543 MJ) over the 12 months, but the cost of gas over that period was only 55% of the cost of electricity. In my case, gas is a cheaper source of energy than electricity.

---

[3] Such as schools, hospitals, government offices, and office buildings.

Table 1.1 My gas and electric bills

Month	Avg. daily temp, °F	Electricity use, kWh	Gas use, Therms	Energy use, MJ			Cost			
				Electricity	Gas	Gas/Electricity	Electricity	Gas	Total	Gas/Electricity
June	74	407	10	1465	1055	0.72	$45.69	$17.58	$63.27	38%
July	75	428	11	1541	1161	0.75	$47.95	$18.68	$66.63	39%
August	70	379	9	1364	950	0.70	$44.05	$17.76	$61.81	40%
September	63	381	9	1371	950	0.69	$44.54	$17.80	$62.34	40%
October	58	396	10	1425	1055	0.74	$47.15	$18.36	$65.51	39%
November	43	654	58	2354	6119	2.60	$72.77	$41.14	$113.91	57%
December	30	747	119	2689	12,555	4.67	$81.26	$72.25	$153.51	89%
January	34	533	73	1919	7702	4.01	$55.77	$46.09	$101.86	83%
February	46	511	31	1839	3271	1.78	$53.78	$27.52	$81.30	51%
March	44	421	36	1515	3798	2.51	$46.01	$29.24	$75.25	64%
April	48	466	30	1677	3165	1.89	$51.24	$24.96	$76.20	49%
May	58	384	18	1382	1899	1.37	$43.53	$20.34	$63.87	47%
Total		5707	414	20,543	43,677	2.13	$633.74	$351.72	$985.46	55%

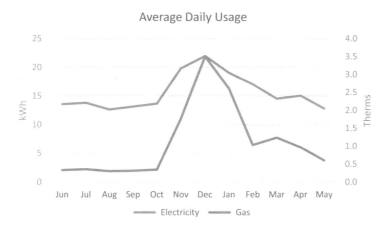

Fig. 1.1 House energy use by month

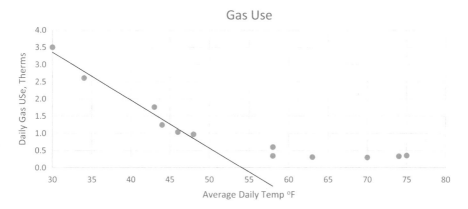

Fig. 1.2 Gas use and outside temperature

It is interesting to look at my energy use by month, which is shown in Fig. 1.1. As you would expect, my gas usage was higher in the winter months than in the summer months. My house uses more energy for heating in the winter than in the summer. In fact, my average daily gas use in December was 13 times as big as in August. You might wonder why the house uses any gas at all during the summer months. There are two reasons: for heating water (since we do not like cold showers) and for cooking.

Figure 1.2 shows how my gas use varied with respect to the average daily outdoor temperature. In each month with temperatures above 58 °F, the daily gas use was approximately constant. But for temperatures below 50 °F, the gas use increased approximately linearly as the average outdoor temperature decreased. The colder the outdoor temperature, the more energy is needed to keep the house at a comfortable temperature, since colder outdoor temperatures lead to more heat loss from the house.

We like a warm house. Any temperature below 70 °F feels cool, so it may seem curious that the gas use does not increase when average outdoor temperature drops below 70 °F. Outdoor temperature fluctuation during the course of the day provides an explanation. The temperature in Colorado typically swings by about 30 °F or 40 °F over the course of a day. So the house warms up during the day when the sun is shining and the outside temperature is higher. The house stores heat in its walls and floors. This heat keeps the house warm in the evening, even without the use of natural gas or electricity for heating.

Figure 1.1 shows that not only gas use but also electricity use varies over the course of the year being higher during the winter months. Can you figure out the reason? There are fewer daylight hours during the winter, so more lights are on during the winter months. The lights use electricity, so electricity use is higher in the winter. The appliances in the house, i.e., the refrigerator, washing machine, drier, and oven, as well as electronic equipment such as the television and computers also use electricity, which explains the energy use during the summer months. However, the swing between summer and winter electricity usage is much smaller than that for gas. The average daily electricity use in December was only 1.7 times as high as that in August, compared with a 13-times swing for gas over the course of the year. This makes it clear that energy used for lighting is relatively small compared to energy used for space heating of the house. Table 1.1 shows that almost five times as much energy was used during December in the form of natural gas as that used by electricity.

It is interesting to compare the energy consumption in my house with the energy that is stored in the gas tank of a car.[4] A full gas tank stores about 1800 MJ of energy.[5] This is more than my house's electricity consumption in most months, and also higher than the gas consumption in the spring and summer.[6] But the rate of energy use is quite different. The house energy consumption is per month, but you probably fill the gas tank in your car multiple times per month.

Energy use in your home will also fluctuate over the course of the year. Unlike me, your electricity use may be highest in the summer if you have air-conditioning. Air-conditioning is energy intensive. Also, your relative costs for gas and electricity may be quite different. Colorado has large supplies of natural gas, so it is relatively cheap here compared to the northeastern USA.

---

[4] It is rather confusing that both natural gas supplied to the house and gasoline are both called "gas." Of course they are not the same. One is liquid and the other is gas.

[5] Detail is provided in Table 2.5, which compares energy storage in an electric vehicle with that in an internal combustion car.

[6] My house's natural gas consumption is much higher than the fuel tank energy in the winter months.

1.3   Energy and the Environment

There is a lot of information about energy and the environment. It is a crucial topic since the environment is our global home. Degradation of the environment can create huge challenges for people, animals, insects, plants, and trees. It is well known that combustion of fossil fuels such as gasoline, natural gas, jet fuel, and coal results in emission of carbon dioxide which traps heat in the planet. There are other environmental issues also. Nuclear power creates nuclear waste which will be radioactive for hundreds of thousands of years. Radioactivity is harmful for people, animals, insects, plants, and trees. Dams that are built in order to generate hydroelectricity affect stream flows and can limit migration of fish. The lakes behind such dams can flood archaeological sites or homes. Wind turbines used to generate wind power kill bats and raptors such as eagles. Industrial scale solar facilities cover large areas with adverse consequences for the habitat.

Why do we tolerate these adverse environmental impacts? The answer is simple. Because energy provides enormous benefits for human life. A much harder issue to resolve is the determination of the optimal trade-off between the benefits and the costs associated with energy. Unfortunately many discussions of this trade-off are motivated by biases or political agendas, and do not focus on the science behind the costs and benefits. As you read this book, you will become aware of the technologies that are used for various forms of energy, what types of energy are used for which purposes, and the fundamentals about how and why energy sources have environmental implications. Please take these learnings, combine them with information about environmental impacts from other sources, and use your knowledge to have intelligent informed discussions about energy and the environment in order to make the world a better place for us and for future generations.

1.4   Summary

This introductory chapter shows how energy use is pervasive in our daily lives. It illustrates this fact in two ways. It tells a story about a daily routine starting with waking up in the morning. The story shows the many ways in which we use energy, probably without really thinking about it. It also uses my utility bills for gas and electricity over the course of a year to show how energy use varies over time. The bills show the relative amounts of gas and electricity that we use in the house. These two illustrations provide human-scale indications of how people use energy. This is important, but so is the industrial scale on which energy is produced and delivered to residential, commercial, and industrial customers, as well as that used for transportation. The chapter also introduces the reality that all forms of energy have environmental impacts. It would be easy to limit these environmental impacts if we limit our consumption of energy. But energy provides such a range of benefits that it would require significant changes in our lifestyle to limit that consumption. This chapter sets the stage for the remainder of the book that explains how energy is used and the technologies behind the various energy sectors (oil, gas, coal, electricity, and petrochemicals).

1.5   Questions

It is wonderful that you are reading this book. It demonstrates that you have an interest in the hugely important topic of energy. Each chapter of the book has questions that will assist you in learning about the details of each of the energy sectors. There are also big picture questions which warrant consideration. This chapter closes with some of those. Maybe you will find time to think about them and to discuss them with your friends. Hopefully, you will return to them when you have read the book. I wonder if your answers will be the same.

1. Will we run out of oil?
2. Will there be an environmental catastrophe due to carbon dioxide?
3. Is fracking bad?
4. Should you turn off all electrical appliances when not in use?
5. What energy conservation measures make sense?
6. What should the role of renewable energy be?
7. Do energy companies have market power?
8. What energy problems do we face now?
9. What energy problems will future generations face?
10. How will the price of energy change in the future?
11. How should energy production and use be regulated?

Fundamentals of Energy

2

2.1 Types of Energy

This book provides a utilitarian perspective on energy by explaining how it is used and the technologies that are involved in various energy sectors. But what is energy? At a fundamental level energy has dual aspects. A definition of energy is the ability to do work or generate heat. Work has a specific meaning in physics, quite different from a number of hours of labor. Formally, work is defined as the product of force times distance moved. Heat is quite different. It is the flow of energy from a hotter object to a cooler one. This book does not focus on formal definitions. Instead it considers the practical implications of energy. For example, each of these activities uses energy:

- Moving an item
- Lifting an item
- Deforming and machining an item
- Heating an item
- Cooling an item
- Melting from solid to liquid
- Boiling from liquid to vapor
- Some chemical processes (e.g., manufacture of ammonia, which is an important agricultural chemical)
- Providing light

There are many forms of energy. Some common forms are shown in Table 2.1.

Let us make this real. It requires more effort to hike uphill than downhill. When you hike uphill you are increasing your elevation and raising your potential energy. You must supply energy in order to increase your potential energy, so it requires effort. On the other hand, you can roll downhill and lower your potential energy. As you roll, you are in motion and as a result you have kinetic energy. Your potential energy is converted to kinetic energy as you roll.

© The Author(s), under exclusive license to Springer Nature Switzerland AG 2021
M. Cronshaw, *Energy in Perspective*,
https://doi.org/10.1007/978-3-030-63541-1_2

Table 2.1 Forms of energy

Potential energy	Energy due to elevation
	Energy available from compression or extension of an object
Kinetic energy	Energy due to motion
Electrical energy	Energy available from electric charge
Sensible energy	Energy due to temperature
Latent heat	Energy intrinsic in a phase (such as liquid or gas)
Chemical energy	Energy available from a chemical reaction
Nuclear energy	Energy available from nuclear fission or fusion
Radiative energy	Energy transferred by electromagnetic radiation

The kinetic energy of an object with mass m and velocity v is $\frac{1}{2}\,m\,v^2$.[1] So kinetic energy increases with the square of the velocity. The potential energy of an object with mass m and height h is $m\,g\,h$, where g is the acceleration due to gravity. So potential energy increases linearly with height.

Here are other examples:

- It takes effort to compress or extend a spring. The compressed or stretched spring contains potential energy that is supplied when the spring is released.
- You may have had a shock from electrical energy. Electricity or burning of a fuel like natural gas can heat something. Once it is heated to a higher temperature, it can release heat to its surroundings when it cools down. It also takes energy to boil water, changing it from a liquid to a gas. This energy is known as the latent heat. Conversely, energy is released when a gas changes to a liquid or when a liquid changes to a solid.
- You release chemical energy when you light a match or a lighter.
- Nuclear energy has been used for both peaceful and military purposes (think power plants and Hiroshima).
- Our sun is an essential source of radiative energy for sustaining life on our planet. A microwave oven is another source of radiative energy, very useful but not as essential as the sun!

Energy has played and will continue to play a pivotal role in our lives. The fact that energy has the dual aspects of work and heat is really important. The industrial revolution was largely enabled by harnessing the chemical energy in fuels for the purpose of doing work. Electricity improves the quality of our lives. It continues this tradition by converting the energy in fuels or from renewable sources into an energy source that can drive motors. Motors can do work. Alternatively, electricity can be used to generate heat or for cooling in refrigerators and air conditioners. Many compounds provide another source of energy. The chemical energy in gasoline, diesel, jet fuel, and residual fuel oil is widely used for transportation by cars, trucks, planes, and ships. The next chapter describes how energy is used as well as compares energy consumption and sources in various countries. It illustrates how energy has an enormous impact on the quality of life.

---

[1] This is translational kinetic energy. There is also a formula for rotational kinetic energy.

2.2   Thermodynamics

Thermodynamics is a branch of physical science concerned with the relationships between heat, work, and different forms of energy. It is a fascinating topic, but

Energy can be in the form of work or heat. Heat usually flows from a warm body to a cool one, but heat pumps can make heat go in the opposite direction! They are used to heat homes or water. Work is performed by an electric compressor in a heat pump. They are able to move about 3–4 times as much heat as the work that is provided.

Source: Kristoferb at English Wikipedia, CC BY-SA 3.0, https://commons.
wikimedia.org/w/index.php?curid=10795550

mostly beyond the scope of this book. This very brief subsection describes two aspects of thermodynamics. Energy conservation is one key aspect. Energy does not disappear. This may sound too good to be true, but even though energy is conserved, it is almost always converted into less useful forms. For example, a car has a lot of kinetic energy when it is moving. When the driver uses the brakes to slow down the car, its kinetic energy decreases. With typical braking, the kinetic energy is dissipated in the form of heat in the brake pads or rotors. This heat is wasted, simply flowing to the environment. Electric cars are not as wasteful when they brake. They use electromagnetic braking to capture much of the kinetic energy and store it in the car's batteries.

Internal combustion engines in cars and trucks also conserve energy, but much of the chemical energy in the fuel does not deliver useful work in the form of kinetic energy (moving) or potential energy (going uphill). Rather, a significant amount of the chemical energy is wasted in the form of heat. You have probably experienced this if you lift the hood on your car after it has been running. Cars have radiators to dissipate this heat. Electrical power plants, discussed later in the book, also waste a large part of the energy from their fuel in the form of heat.

You may think that there would be a scientific solution for this wasted heat energy. However, thermodynamics shows that energy is inevitably wasted when heat is converted to work. The conversion of heat to work is arguably one of the most important inventions ever. Significant breakthroughs in this conversion occurred in the eighteenth century, with the invention and improvement of the steam engine. Steam engines powered by coal enabled rail transport and the use of machines for manufacturing. The invention was so significant that it led to the industrial revolution. Today's cars, trucks, and trains are no longer fueled by coal. Instead most of them use gasoline or diesel as fuel. But many of the fundamentals are the same. The fuel is burned to generate heat, and that heat is converted to work (i.e., motion).

2.3   Energy and Power

As mentioned above, energy is the ability to do work or to generate heat. For example, fuels have chemical energy. Similarly a hot item has thermal energy. Power is related to energy, but it is a different concept. Power is the rate at which energy is used. For example, the natural gas used in a stove top has chemical energy. This energy is released when the gas burns. When the burner is turned down the flow rate of gas is low, it is burned slowly, and the power in the gas flame is low. Conversely the power is high when the burner is turned up, and the flow rate of gas is high.

Table 2.2 shows measurement units that are used to define amounts of energy.[2] The table has several different units. Each one is in common use. Two of the units are related to temperature increase. One of them, the joule, has two alternative

Table 2.2 Energy units

Unit	Symbol	Definition
Foot-pound (imperial)	Ft-lb	Amount of work required to move one pound by one foot
Newton[a]-meter (SI[b] unit)	N-m	Amount of work when applying a force of one newton over one meter
Joule (SI unit)	J	Energy transferred to an object when it is moved a distance of one meter in the direction of a force of one newton
		Energy dissipated as heat when an electric current of one ampere flows through a resistance of one ohm for one second
Calorie	Cal	Energy required to raise the temperature of 1 gram of water by 1 °C
British thermal unit (imperial)	BTU	Energy required to raise the temperature of 1 pound of water by 1 °F
Quad		10^{15} BTU
Watt-hour	Wh	Energy when using one joule per second for one hour

[a]One newton in the force that accelerates one kilogram by one meter per second squared
[b]SI is the Système International, commonly known as the metric system

---

[2]The appendix to this chapter has a discussion of many other types of units, such as length and area, which are used in energy sectors.

Table 2.3 Prefixes for units

Prefix	Pronunciation	Meaning
k	kilo	Thousand in SI
M	Em	Thousand in imperial; million in SI
MM	Em Em	Million in imperial
G	Giga	10^9
T	Tera	10^{12}
P	Peta	10^{15}
E	Exa	10^{18}
Z	Zeta	10^{21}

Table 2.4 Units for power

Unit	Symbol	Definition
Watt	W	1 J/s
Horsepower	Hp	Energy expended when lifting 33,000 pounds by 1 ft in 1 min[a]

[a]https://www.greentechmedia.com/articles/read/energy-and-power-units-the-basics-8#gs.9savc3

definitions: one in terms of work (force × distance) and the other in terms of electrical current.

> Many of the units are named after distinguished scientists. Isaac Newton is known for his laws of motion and insights about gravity. James Prescott Joule was an English physicist in the nineteenth century. James Watt commercialized the steam engine.

The units often have prefixes to denote multiples of the unit itself. Some common prefixes are shown in Table 2.3. For example, MMBTU means one million BTUs, GJ means one billion joules, and GWh means one billion watt-hours. As another example, if you use a 100 watt light bulb for 1 hour, it consumes 100 Wh or 0.1 kWh.[3]

Power is the rate of energy usage. There are two common units for power as shown in Table 2.4.

2.4 Energy Storage

It is really convenient to have supplies of useful things on hand. Think of your food cupboard or your bank account. Energy is a useful thing, so it is great to have a store of it. Electrical batteries store energy, so our mobile phones, flashlights, and other

---

[3] Some of this energy is provided as light, but most of the energy from incandescent bulbs is in the form of heat rather than light. If the retail price of electricity is $0.15 per kWh, then it costs $0.015, i.e., 1.5 cents, to operate a 100 W bulb for 1 hour.

small appliances function when we need them. However, the amount of energy that can be stored in batteries is unfortunately small. Consequently considerable design effort is devoted to reducing the energy consumption of small appliances like cellphones. Much research is being undertaken to increase the storage capacity of batteries.

Tesla installed a huge battery system in Australia in 2018 (See Fig. 2.1). The battery uses lithium-ion technology, as do batteries in cellphones and Tesla cars. The Australian battery system can store 129 MWh, and can discharge up to 100 MW. To put this in perspective, an AA alkaline battery stores about 0.000004 MWh, and large power plants have generation capacities of about 1000 MW. So the battery system stores the equivalent of about 25 million AA batteries and can generate about one-tenth of the power of a large power plant.
Source: https://electrek.co/2018/09/24/tesla-powerpack-battery-australia-cost-revenue/

Fig. 2.1 Tesla's Australian battery system

There are very few known technologies for storing large amounts of energy, except for the energy content of hydrocarbon or nuclear fuels. One nonfuel method for storing large amounts of energy is pump storage. A pump storage system consists of two water reservoirs, one above the other. Energy is stored by pumping water from the lower reservoir to the upper reservoir. This increases the potential energy of the water. Energy is recovered by allowing the water to flow from the upper reservoir to the lower reservoir through a pipe. The dropping water turns a turbine which is connected to an electric generator. Part of a pump storage system is shown in Fig. 2.2.

Fig. 2.2 A pump storage reservoir

The Cabin Creek generating station is a pump storage facility near Georgetown, Colorado. The system can store about 1318 MWh and can generate up to 324 MW.

Flywheels are another method for storing energy. They contain a heavy wheel that spins. The spinning wheel stores energy in the form of rotational energy. Flywheels usually operate in a vacuum with low-friction bearings in order to reduce frictional losses. Today's flywheels can store between 3 kWh and 133 kWh.[4] One of their advantages is that they can be charged up relatively quickly, by increasing the angular velocity of the spinning wheel. They have been used in some cars and buses. Kodiak, Alaska, has an unusual application of a flywheel. Much of the energy supply in Kodiak is provided by wind, which is intermittent. A big Kodiak energy consumer is an electric dock crane that draws a significant amount of energy. The crane uses substantial amounts of power when it is lifting and generates power when it is lowering something. The Kodiak flywheel can respond quickly, supplying power to the electrical grid for the crane when needed, and storing it when power is being generated by the crane.[5]

---

[4] https://en.wikipedia.org/wiki/Flywheel_energy_storage

[5] http://acep.uaf.edu/acep-news/2018/23-july-acep-this-week/kodiak%E2%80%99s-advanced-microgrid-system-includes-flywheel-energy-storage.aspx

Table 2.5 Car energy storage

	Tesla model S AWD-85D	Car	Tesla/car
Battery capacity, kWh	85		
Fuel tank capacity, US gallon		15	
Gasoline heat content, BTU/US gallon		115,000	
Fuel tank capacity, MMBTU		1.73	
Energy storage, MJ	306	1,820	0.17
Range, miles	265	405	0.65
Miles per MJ	0.87	0.22	3.89
Miles per gallon gasoline equivalent	105	27	3.89
Storage system weight, kg	544[a]	41.6	13.1
Energy storage density, MJ/kg	0.56	43.8	0.013

[a] http://www.roperld.com/science/TeslaModelS.htm

Liquid fuels are a great way to store large amounts of energy. We illustrate this by comparing energy storage in electric cars with that in gasoline tanks of conventional cars. Electric cars store their energy in batteries, while traditional cars store energy as gasoline or diesel in a fuel tank. Table 2.5 compares the energy storage in the two types of vehicles. The table shows that a Tesla model S electric car stores only about 1/6 as much energy as a traditional gasoline-powered car, yet its range is about 2/3 that of a gasoline-powered car. This demonstrates that the Tesla uses energy more efficiently than a gasoline-powered car. In fact, the Tesla is able to drive almost four times as many miles per unit of stored energy as a gasoline-powered car. Clearly the Tesla design has taken energy efficiency very seriously. A gasoline-powered car stores a large quantity of energy in its fuel tank, but it does not use that energy very efficiently.

The table shows another key feature about energy storage, namely energy density. The batteries in the Tesla weigh about 13 times as much as the weight of fuel in a car but store only 1/6 of the energy. So the energy stored in the Tesla batteries per kilogram of battery weight is only 1/80 of the energy content of gasoline per kilogram. The energy used by a car depends on several things including its weight. Since the Tesla batteries weigh so much more than the weight of fuel in a traditional car, the Tesla must use its stored energy much more efficiently.

2.5   Hydrocarbons

2.5.1   Chemistry

Oil, natural gas, and coal are all mixtures of hydrocarbons. As their name indicates, hydrocarbons consist of hydrogen and carbon. They may also contain small amounts of other elements such as oxygen or sulfur. Carbon is tetravalent, meaning that each carbon atom has four bonds in any molecule. The chemical bonds are formed by electrons. The simplest hydrocarbon is methane, which is the main constituent of natural gas. Its chemical formula is CH_4. The "C" stands for carbon, and the "H" for hydrogen. The chemical formula shows that each molecule of methane has one carbon atom and four hydrogen atoms. Its chemical structure is shown in Fig. 2.3.

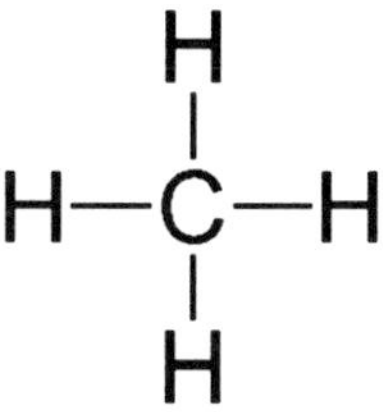

Fig. 2.3 Chemical structure of methane

The figure shows the four bonds emanating from the carbon atom and the four hydrogen atoms, each of which is univalent, i.e., has one bond per atom. There are many other hydrocarbons besides methane. Several are shown in Fig. 2.4.

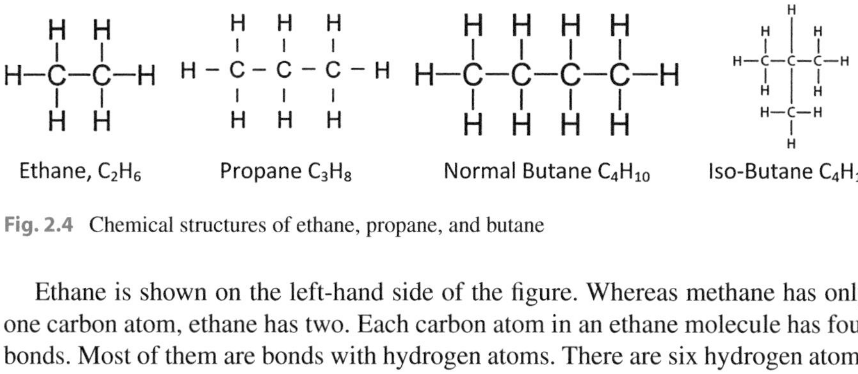

Ethane, C_2H_6 Propane C_3H_8 Normal Butane C_4H_{10} Iso-Butane C_4H_{10}

Fig. 2.4 Chemical structures of ethane, propane, and butane

Ethane is shown on the left-hand side of the figure. Whereas methane has only one carbon atom, ethane has two. Each carbon atom in an ethane molecule has four bonds. Most of them are bonds with hydrogen atoms. There are six hydrogen atoms in each molecule of ethane. The figure also shows propane with its three carbon atoms, as well as two forms of butane, each with four carbon atoms. Normal butane has all four carbon atoms in a line, whereas isobutane is a branched hydrocarbon with one of its carbon atoms out of line with the other three. Both normal butane and isobutane have the same chemical formula, but they have different physical properties. They are known as isomers of butane. A numbering scheme is used to identify the location of branching when it occurs. For example, isobutane has a chain of three carbon atoms with a fourth carbon atom connected to the second carbon atom in the three-carbon chain. The CH_3 group that is attached to the second carbon atom is known as a methyl group since it is similar to methane. Isobutane can be referred to as 2-methylpropane, to show that a methyl group is attached to a three-carbon chain (i.e., like propane), and the "2" shows that the methyl group is on the second carbon in the three-carbon chain. Figure 2.5 shows two alternative representations of another branched hydrocarbon, 2-methylbutane. The "2" shows that a methyl group is on the second carbon in a four-carbon chain (butane), which is most easily seen on the left-hand panel in the figure. This compound has five carbon atoms. The right panel in the figure shows a shorthand depiction of the same molecule. Some of the carbon atoms are specifically shown, while others are inferred to be present at the kinks in the line in the figure. There are three explicit "C"s and two kinks, and hence five carbon atoms.

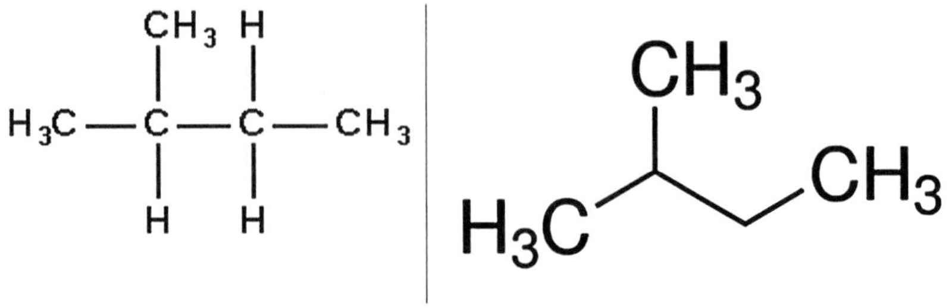

Fig. 2.5 Two alternative representations of 2-methylbutane

Some hydrocarbon molecules have rings of carbon atoms with something other than the single bonds shown above. They have so-called resonance bonds. Hydrocarbon molecules with resonance bonds are called "aromatic." Benzene is an example of an aromatic molecule. It is discussed below. Hydrocarbons that do not have rings of carbon atoms with resonance bonds are called "aliphatic" hydrocarbons.

Hydrocarbons that contain only single bonds are said to be "saturated," and are called alkanes or paraffins. There are many other alkanes besides those shown in the figures above. Pentane has five carbon atoms, hexane six, heptane seven, and octane eight. All aliphatic hydrocarbons with at least four carbon atoms have isomers, some with straight chains and some branched. The general formula for aliphatic hydrocarbons is $C_n H_{2n+2}$ where n is a positive integer, e.g., 1, 2, 3, and 4. Methane corresponds to $n = 1$, ethane to $n = 2$, propane to $n = 3$, and so on. At atmospheric conditions, hydrocarbons with n between 1 and 4 are gases, while those with n between 5 and 17 are liquids, and those with n above 17 are colorless waxlike solids.[6] The chemical composition has important implications for emissions when they are burned. This is discussed below.

There is another class of hydrocarbons known as alkenes or olefins. Ethylene is an example with two carbon atoms, and is shown on the left side of Fig. 2.6. The "-"notation in the figure means a single bond, while the "=" notation means a double bond.

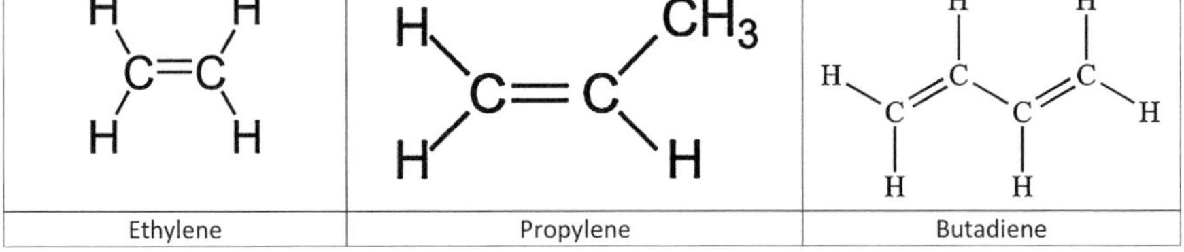

| Ethylene | Propylene | Butadiene |

Fig. 2.6 Chemical structure of three alkenes

Ethylene has two carbon atoms, the same as ethane, but it has a double bond between the carbon atoms rather than a single bond. Each carbon atom still has four bonds: two single bonds to hydrogen atoms and one double bond between the two

---

[6] http://petrowiki.org/Crude_oil_characterization?rel=1

carbon atoms, for a total of four bonds. Ethylene has two fewer hydrogen atoms per molecule than ethane. There are many other alkenes. For example, propylene (shown in the middle of the figure) has three carbon atoms. Two of them have one double bond and two single bonds. The third carbon atom has only single bonds. The right panel of the figure shows another alkene, butadiene, which has four carbon items and two double bonds. It is an important chemical for the manufacture of synthetic rubber.

Alkenes do not occur naturally in significant quantities but can be manufactured from alkanes using a catalyst and a high-temperature reaction. This reaction releases hydrogen gas. Alkenes are important as precursors for the manufacture of polymers, commonly known as plastics, discussed in Chap. 8. Plastic grocery bags are polyethylene. Polypropylene, the polymer made from propylene, is used in a wide variety of applications including packaging, furniture, and underwear. Most rubber that is used nowadays is polybutadiene, which is made from butadiene.

For completeness, there is yet another class of hydrocarbons known as alkynes, in which there is a triple bond between carbon atoms. Figure 2.7 shows the chemical structure of acetylene, which has two carbon atoms, and a triple bond between them. Like ethane, acetylene has two carbon atoms per molecule. Ethane, ethylene, and acetylene are each gases at atmospheric conditions. Acetylene is used for welding since the temperature of its flame when burned with air is very high. Hydrocarbons with double or triple bonds are called "unsaturated" in contrast to hydrocarbons that only have single bonds.

Fig. 2.7 Chemical
structure of acetylene

$$H-C\equiv C-H$$

There are also hydrocarbons that contain rings of carbon atoms. For example, cyclohexane contains a ring with six carbon atoms and is shown on the left of Fig. 2.8. It contains only single bonds, and so is aliphatic.

There are also molecules with rings that have some double bonds. The central image in Fig. 2.8 was developed by a German chemist August Kekulé. It shows a ringed compound called benzene, which has three double bonds and three single bonds. Each carbon atom has four bonds in this configuration. More modern representations of benzene are as shown at the right of the figure, in which the double-bond electrons are uniformly distributed around the carbon ring as so-called pi electrons, but the chemical composition is the same, containing six carbon atoms and six hydrogen atoms. Hydrocarbons with pi electrons are called aromatic.

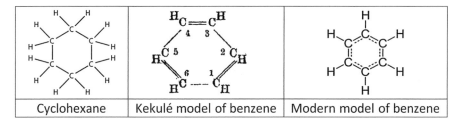

| Cyclohexane | Kekulé model of benzene | Modern model of benzene |

Fig. 2.8 Chemical structure of two carbon molecules with rings

The preceding discussion has covered relatively simple hydrocarbons. Figure 2.9 shows an example of the complex chemical structure of coal.

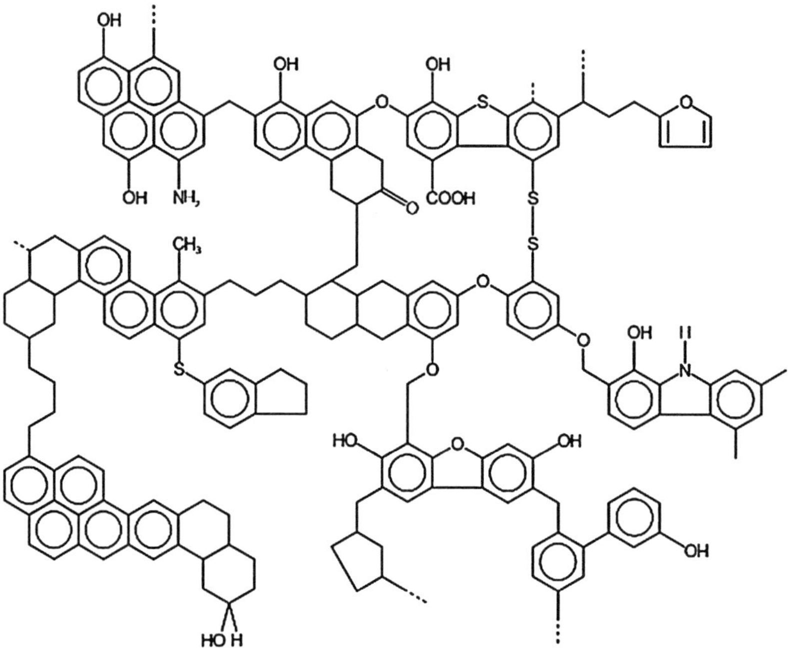

Fig. 2.9 Example chemical structure of coal

For clarity, the letters "C" and "H" are mostly absent from the figure. The hexagons with circles are like the benzene ring of Fig. 2.8, while the hexagons without rings are like cyclohexane in that figure. The compound also contains sulfur (S), oxygen (O), and nitrogen (N). The structure also includes some rings with five carbon atoms.

This section has introduced many types of hydrocarbons. The various types are summarized in Table 2.6. Despite the complexity, the most important takeaway is that hydrocarbons consist primarily of hydrogen and carbon, and that the ratio of

Table 2.6 Summary of hydrocarbon types

	No rings		Rings
	Aliphatic		*Aromatic*
Saturated (only single bonds)	Alkanes, paraffins	Cycloalkanes Naphthenes	NA
Unsaturated (some double or triple bonds)	Alkenes, alkynes, olefins	Not common	Benzene, coal, etc.

carbon to hydrogen in a hydrocarbon molecule depends on the specific hydrocarbon. For example, methane has four hydrogen atoms for each carbon atom, while octane, C_8H_{18}, has 2.25 hydrogen atoms per carbon atom, and benzene, C_6H_6, has only one hydrogen atom per carbon atom.

2.5.2 Gas

Natural gas produced from subsurface reservoirs is a mixture of hydrocarbons, of which methane is the main component. Chapter 5 provides details about exploring for and producing natural gas. In common usage the word "gas" refers to gasoline, which is a liquid fuel used for cars, trucks, lawn mowers, boats, and so on. This is not the same as natural gas, which is not a liquid but rather a gas at atmospheric conditions. There are two main classifications of gas: associated gas is produced from wells that primarily produce oil together with some gas, while nonassociated gas is produced from wells that primarily produce natural gas and may also produce some hydrocarbon liquids. Somewhat confusingly, the liquids that are produced from a gas well are often referred to as natural gasoline. Alternatively the liquids may be called condensate or natural gas liquids (NGLs).

Natural gas often contains ethane, propane, butane, and pentane together with methane. It may also contain carbon dioxide, nitrogen, hydrogen, helium, and/or hydrogen sulfide. Hydrogen sulfide is extremely toxic to humans and animals. Gas with hydrogen sulfide is referred to as sour gas. It requires special handling due to its toxicity.

Since gas in an underground reservoir is in contact with water in the reservoir, natural gas also contains water vapor. Water vapor, carbon dioxide, and nitrogen have no value as a fuel. Furthermore water vapor may condense into liquid water

> The Tengiz field in Kazakhstan is located in the Caspian Sea. It produces oil and associated gas. Pressures and temperatures at the surface of the wells are extremely high. Both the oil and gas have high sulfur contents. The high pressure, high temperature, and sulfur content have required the use of special materials for wells and surface facilities.

during downstream handling and transportation. So natural gas from subsurface reservoirs is processed in the vicinity of gas wells to remove water, hydrogen sulfide, carbon dioxide, and other inert components. Some gas reservoirs also contain small amounts of hydrogen and helium as well as methane, the main compound. Hydrogen and helium are valuable gases but are only available in a few gas wells. These high-value products are extracted from natural gas and sold separately.

2.5.3   Oil

Oil is a liquid that is produced from subsurface reservoirs. It is referred to as crude oil. Like gas, it is a mixture of hydrocarbons. Oil wells also produce natural gas, referred to as associated gas. Crude oil contains such a large number of different hydrocarbons that the nature of different crude oils is often summarized by only few parameters. One of these is whether the crude oil contains sulfur in the form of hydrogen sulfide or other compounds (known as mercaptans). Oil that contains sulfur is called sour.

A common parameter for classifying crude oil is its API gravity. API gravity is a measure of the density of the crude oil. The API gravity of crude oils ranges from about 10 to 45. A low API gravity corresponds to heavy or dense oil; a high API gravity corresponds to light oil. Specific gravity is a common measure of the density of a liquid. Water has a specific gravity of 1.0. Liquids with a specific gravity less than 1.0 are less dense than water, and therefore float on water. Liquids with a specific gravity greater than 1.0 are denser than water, and therefore sink in water. The relationship between specific gravity (SG) and API gravity is as follows:

$$SG = \frac{141.5}{131.5 + API}.$$

Note that the specific gravity is 1.0 if the API gravity is 10, and it is less than 1.0 if the API gravity is greater than 10.

> Many oil and vinegar salad dressings have an upper layer and a lower layer. Vinegar is mostly water. Oil, being less dense than water, floats on the vinegar. The two liquids can be made to mix together in an emulsion by shaking or by adding something else like mustard. Oil and water that are produced from subsurface wells often form an emulsion. Purchasers do not want the water, so various methods with equipment and heat are used to break the emulsion.

Figure 2.10 shows the API gravity and sulfur content of various types of crude oil from different worldwide locations. West Texas Intermediate (WTI) is a categorization for crude oil from West Texas. Brent Blend is a categorization for crude oil from the North Sea between Scotland, Norway, and the Netherlands. Both of these are light crude oils with an API gravity of about 39. Maya crude oil, from Mexico, is heavier (lower API) and has more sulfur.

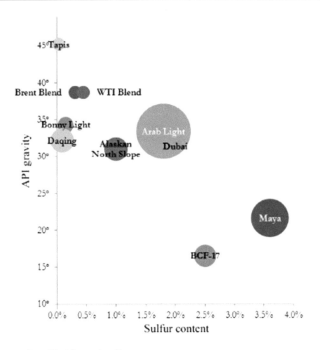

Fig. 2.10   Types of worldwide crude oil

A list of some of the hydrocarbons that are present in crude oil from an Oklahoma field is shown in Table 2.7. The left column lists paraffins which are saturated hydrocarbons. Normal paraffins have only straight chains. Compounds such as iso-butane or 2-methylpentane are not straight chains, but have branches instead. Naphthenes are also saturated, but have rings of carbon atoms. For example, methyl cyclohexane has a ring of six carbon atoms with a methyl group attached to one of the carbon atoms in the ring, as shown in the middle of Fig. 2.11. Aromatics also have carbon rings, but they are not saturated, i.e., they have double bonds or a structure with a benzene ring (see Fig. 2.8).

Table 2.7   Compounds present in Oklahoma crude oil

Paraffins	Naphthenes	Aromatics
All normal paraffins to $C_{10}H_{22}$	Cyclopentane	Benzene
Isobutane	Cyclohexane	Toluene
2-Methylbutane	Methylcyclopentane	Ethylbenzene
2,3-Dimethylbutane	1,1-Dimethyl cyclopentane	Xylene
2-Methylpentane	Methylcyclohexane	1,2,4-Trimethylbenzene
3-Methylpentane	1,3-Dtmethylcyclohexane	
2-Methylhexane	1,2,4-Trimethylcydohexane	
3-Methylhexane		
2-Methylheptane		
2,6-Dimethylheptane		
2-Methyloctane		

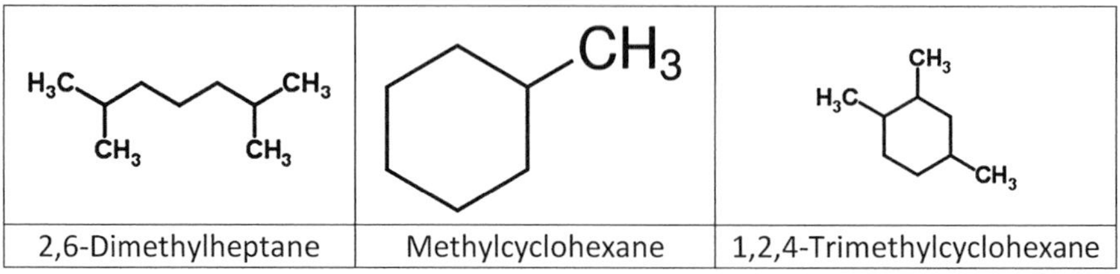

2,6-Dimethylheptane	Methylcyclohexane	1,2,4-Trimethylcyclohexane

Fig. 2.11 Chemical structure of three crude oil compounds

The structure of some of the hydrocarbons listed in the table is shown in Fig. 2.11. Note how "cyclo" in the compound name indicates carbon rings, and note also that sometimes the figures show carbon atoms explicitly with a "C" and sometimes they indicate carbon atoms simply as a vertex (i.e., kink or bend).

Paraffins can cause problems in oil wells. They have a tendency to become solid, resulting in a narrowing of the region available for flow in wells and pipes. In fact many candles that are used in homes are made from paraffin wax. They work well as candles, since they are solid flammable hydrocarbons, but they are a nuisance in oil wells, sometimes completely blocking flow in a well.

2.5.4   Coal

There are four types of coal as shown in Table 2.8. The various types are known as "ranks."

Table 2.8 Ranks of coal

Rank	Description	Picture	Heat content, MJ/kg
Anthracite	Hard, brittle, black, lustrous. Known as hard coal		32–33
Bituminous	Shiny, smooth. High heating value		23–33
Subbituminous	Black, dull color		18–23
Lignite	Brown coal. Least concentration of carbon. Lowest grade		17–18

Anthracite has the highest heating value and lignite has the lowest heating value. Coal can contain various impurities including sulfur and mercury. Most coal is used for generating electricity, but some is used in the manufacture of steel.

When coal is burned, any sulfur forms sulfur dioxide. If this is emitted as a combustion product, it reacts with water in the atmosphere forming sulfuric acid. This falls to earth as rain and can result in acidification of rivers, streams, and lakes. Sulfur dioxide can be removed from the exhaust stream by passing the combustion gases through a flue gas desulfurization (FGD) unit. Mercury is a toxic metal. Emission of mercury from coal combustion is regulated in the United States in order to reduce the adverse health impacts.

2.5.5 Combustion and Emissions

Combustion is the name for burning fuels. Generally combustion requires both a fuel and oxygen.[7] Conveniently air is a mixture of gases that contains about 21% oxygen. It also contains about 78% nitrogen, 1% argon, and small amounts of other gases. Combustion of hydrocarbons is the main source of energy provided by them. For example, combustion occurs in the burner of a gas stove, in gas-fired water heaters, in the cylinders of car and truck internal combustion engines, in jet engines, in a fireplace, and also in many types of electrical power plant.

During combustion with air, fuel is consumed in a chemical reaction with the oxygen from the air, and combustion products are formed. In the case of hydrocarbons, the combustion products are carbon dioxide (CO_2) and water (H_2O). If combustion is incomplete, then carbon monoxide (CO) may also be produced. A general chemical formula for hydrocarbons is C_mH_n, where m and n are integers. If a hydrocarbon is burned completely then all of the carbon and hydrogen in the fuel are converted to carbon dioxide and water. The chemical reaction for combustion with oxygen is written as[8].

$$C_mH_n + (m + n/4)\ O_2 = m\ CO_2 + (n/2)\ H_2O.$$

The left-hand side represents the combination of hydrocarbon and oxygen, while the right-hand side represents the combustion products. Notice that there are m carbon atoms on each side of the equation, n hydrogen atoms on each side, and also the same number of oxygen atoms on each side.[9]

Figure 2.12 shows the chemical structures of carbon dioxide and water. The left side of the figure shows carbon dioxide. The carbon atom has two double bonds, which are somewhat analogous to four single bonds. Each oxygen atom has a double bond to the carbon atom. Oxygen is divalent, meaning that it has two bonds in

---

[7] Combustion can occur with other oxidizers instead of oxygen.

[8] The "=" sign in the chemical reaction formula does not mean that both sides are equal in an arithmetic sense. It means that the compounds on the left side of the "=" sign react to form the products on the right side. This is the standard usage for any chemical reaction.

[9] (n/2) molecules of water, each of which contains two molecules of hydrogen.

$$O = C = O \qquad H\text{-}O\text{-}H$$

Fig. 2.12   Chemical structures of carbon dioxide and water

chemical compounds. This divalence is also shown on the right side of the figure, which shows the chemical structure of water. Each oxygen atom has a single bond to two separate hydrogen atoms. Hydrogen is monovalent, meaning that it has one bond.

Since air contains about four times as much nitrogen as oxygen, the chemical reaction for combustion with air is approximately.

$$C_mH_n + (m + n/4)\ O_2 + (4\ m + n)\ N_2 = m\ CO_2 + (n/2)\ H_2O + (4\ m + n)\ N_2.$$

Nitrogen is generally inert, i.e., not reactive. However, since combustion occurs at high temperatures, some nitrogen in the air reacts with oxygen in the air to form nitrogen oxides such as NO_2 or NO. These oxides are sometimes abbreviated as NOx, rhyming with "socks." They contribute to smog and acid rain, and are a major source of air pollution. The brown cloud in some cities is caused by NOx.

The energy released during combustion with oxygen is known as the "heat of combustion." Since combustion occurs at high temperatures, the water produced as a combustion product is in the form of steam, i.e., water vapor. If the combustion products are allowed to cool, and the steam is allowed to condense into liquid water, rather than water vapor (i.e., steam), then the latent heat of vaporization is released. This latent heat is additional energy realized from the combustion reaction. Thus two different values are commonly reported for the heat of combustion: the higher heating value (HHV) and the lower heating value (LHV). The LHV, which does not allow for the cooling, is more representative of the typical energy achieved during actual combustion, since complete cooling of the combustion products does not usually occur. The HHV allows for cooling to a standard temperature and for condensation of the steam. The heats of combustion for various fuels are shown in Table 2.9.

Table 2.9   Heat of combustion for various fuels

Fuel	HHV BTU/lb.	LHV BTU/lb.
Methane	23,900	21,500
Propane	21,700	19,934
Diesel	19,300	
Coal (anthracite)	14,000	

As shown above, carbon dioxide is one of the two reaction products during combustion of hydrocarbons. Coal is another common fuel. It consists mostly of carbon and hydrogen, although it also contains oxygen, sulfur, nitrogen, ash, and water. Consequently the use of hydrocarbons or coal for energy generation results in emissions of carbon dioxide. Table 2.10 shows the emissions of carbon dioxide per MMBTU of energy provided during combustion. Note that combustion of coal produces about twice as much carbon dioxide per MMBTU as combustion of natural gas, which is composed mainly of methane.

Table 2.10 CO$_2$ Emissions for various fuels

Fuel	CO$_2$ emissions lb. per MMBTU
Natural gas	117
Propane	139
Diesel	161
Gasoline	157
Jet fuel	156
Coal (anthracite)	210
Coal (bituminous)	206

There is a relatively simple chemical explanation for the variation in carbon dioxide emissions from the different fuels. During combustion, the chemical bonds in the fuel are broken and new bonds are formed in the combustion products. So for example, during combustion of methane four carbon-hydrogen bonds are broken per molecule of methane, and two carbon-oxygen double bonds are formed per molecule of CO$_2$, as well as two hydrogen-oxygen bonds per molecule of water (H$_2$0). The ratio of hydrogen to carbon in methane is 4:1, whereas it is 8:3, or about 2.67 in propane. The higher ratio in methane means that more of the energy released during combustion of methane comes from breaking carbon-hydrogen bonds, than during combustion of propane, which also involves the breaking of carbon-carbon bonds.

2.6 Summary

This chapter provides a brief overview of some of the science that is relevant for energy. It explains how energy has two aspects: the ability to do work and for heat flow. Work is formally defined as force multiplied by distance. This makes intuitive sense. More energy is required to move further and/or to overcome higher levels of force. The dual nature of energy reveals the important relationship between heat and work. Thermodynamics is the study of this relationship. This chapter provides a very basic and high-level treatment of that important topic, in particular noting that energy is conserved. For example, when an object drops from a higher elevation to a lower one, its potential energy decreases, but its kinetic energy increases as it picks up speed.

The chapter also explains the crucial difference between energy and power. Power is the rate at which energy is transferred. The same amount of energy transferred in a shorter time corresponds to higher levels of power. The gas burner on a stove provides a clear illustration of energy and power. The chemical energy in the gas is released when the gas burns. The power in the flame is high when the burner is turned up high, and low when the burner is turned down low.

Energy storage is also discussed. It is helpful to be able to store anything that is useful. Since energy is useful, it is helpful to be able to store it. Small amounts of energy are stored in the batteries that power small appliances such as cellphones. There are technical issues that limit the amount of energy that can be stored in batteries, although this is an active research area. A low-tech but effective energy storage technology is pumped storage in which water is pumped up to a higher elevation and released to flow downhill when energy is required.

The chapter provides a short introduction to the chemistry of hydrocarbons and the chemical reaction that occurs when hydrocarbons are burned in air. That process is known as combustion. The combustion products are carbon dioxide and water. Impurities in the fuel such as sulfur will also result in other combustion products such as sulfur dioxide. It causes acid rain when released into the atmosphere.

An appendix to the chapter describes the various units that quantify amounts of length, weight, energy, and power, as well as techniques for converting between the different units that are used for the same item.

2.7   Questions

1. Suppose that the retail price of electricity is $0.15 per kWh, and the retail price of diesel for vehicles is $3.50 per gallon. The density of diesel fuel is about 6.9 pounds per gallon. Which source provides energy at a lower cost? Hint: Use HHV in Table 2.9 for diesel, and conversion factors in Table 2.14.

2. A fully charged manganese alkaline AA battery provides 1.5 V and contains about 13,000 J. The voltage is insufficient to power a typical 100 W incandescent lightbulb. But if it could do so, how long would the bulb be lit for?

3. The upper reservoir at the Cabin Creek pumped storage facility is 650 feet above the lower reservoir. The upper and lower reservoirs each store 1977 acre-feet of water.

   (a) There are 1233 cubic meters in an acre-foot. What is the capacity of the reservoirs in cubic meters?

   (b) The density of water is 1000 kg per cubic meter. What is the mass of water in a full reservoir?

   (c) There are 3.28 feet in a meter. How much higher is the upper reservoir in meters?

   (d) The potential energy (in joules) of a mass m (in kg) at a height of h (in meters) is m g h where $g = 9.81$ m s^{-2} is the acceleration due to gravity. What is the potential energy in joules in the upper reservoir when it is full?

   (e) There are 3.6 MJ in a kWh. What is the potential energy of a full upper reservoir in MWh?

4. Explain why the Kodiak electric crane described in Sect. 2.4 consumes energy when lifting a load. Why does it generate energy when it is lowering a load?

5. An LED bulb that puts out as much light as a 60 W incandescent bulb consumes 8.5 W.

   (a) What is the percentage savings in power of an LED bulb compared to a 60 W incandescent bulb?

   (b) If you use a 60 W bulb every day for 4 hours per day on average, and electricity costs $0.10 per kWh, how much money would you save per year by using an LED bulb?

6. WTI and Brent each have API gravity of about 39°. What is the specific gravity of these crude oils? Would they float on water or sink?

7. Draw the chemical structures of 2-methylpentane, 3-methylpentane, and 1,3-dimethylcyclohexane.

8. How many isomers are there of methane, ethane, propane, butane, and pentane?

9. Show that there are same number of oxygen atoms on each side of the combustion equation in Sect. 2.5.5.

10. The molecular weights of carbon and oxygen are 12 and 16, respectively.

   (a) What is the molecular weight of CO_2?
   (b) During complete combustion, each carbon molecule in the fuel produces one molecule of CO_2. Using the molecular weights, determine how many pounds of CO_2 are produced during complete combustion per pound carbon in a fuel?

11. Based on HHV:

   (a) How many pounds of anthracite coal must be burned to produce one million BTU of energy?
   (b) How many pounds of methane must be burned to produce one million BTU of energy?
   (c) Coal consists of about 78% carbon by weight, and methane consists of 75% carbon by weight. Calculate the CO_2 emissions per pound of coal and per pound of methane assuming complete combustion.

12. The English still use stones for measuring body weight. If someone weighs 13 stone and 6 pounds, what is their weight in pounds?

13. The specific gravity of carbon dioxide is greater than one. In 1986 a large eruption of carbon dioxide occurred at Lake Nyos in Africa. Why do you think about 1700 people and 3500 livestock died as a result?

14. If an engine puts out power at a rate of 300 horsepower, how much energy in kWh does it put out during 1 hour? How much energy is this in BTUs?

15. One foot is 0.3048 meters, or 0.3048 m.

   (a) Show that 1 ft^2 is 0.0929 m^2.
   (b) Show that 1 ft^3 is 0.0283 m^3.

Appendix: Units and Unit Conversion

Imagine that you saw an ad offering a pile of carrots for sale for $1. Doubtless you would have a few questions, such as "how old are they?" and "how many are there?" Even if you found out that they were freshly picked, and that there were four carrots, some uncertainty would remain. An obvious open issue is their size or weight. This is where a standard measure of weight is really helpful. We are so used to such an idea that we probably do not give it any thought. But this appendix discusses the idea of units for various items such as weight, length, temperature, energy, and power.

There are many standards for weight, such as ounces, pounds, grams, kilograms, and tons. You may wonder why there are so many. It is mostly due to history. The word ounce is derived from the Latin word *uncia*, which was 1/12 of a Roman pound.[10] The ounce is part of the avoirdupois measurement system of weights, which was first commonly used in the thirteenth century.[11] You can understand why a system of weights would have been used such a long time ago, in order to limit unscrupulous activities by traders!

Today, 16 ounces make up an avoirdupois pound. There are also other units of weight. One such measure is a ton, which is unfortunately ambiguous. A short ton is 2000 pounds, while a long ton is 2240 pounds. A ton is twenty hundredweight, where a hundredweight is 100 pounds in the USA and 112 pounds in the UK. In the UK a hundredweight is eight stone where a stone is 14 pounds. A photograph of a seventeenth-century stone weight is shown in Fig. 2.13. It was used in the wool trade.

Fig. 2.13 Seventeenth-century English stone weight

It is inconvenient to have a measurement system with irregular units, like ounces, pounds, stones, hundredweights, and so on. In 1960 the International System of Units, known as SI, was established. It is a decimal system of units that is easy to work with. The SI defines the seven units shown in Table 2.11 as base units.

Table 2.11 SI base units

Category	Unit	Abbreviation
Time	Second	s
Length	Meter	m
Mass	Kilogram	kg
Electric current	Ampere	A
Temperature	Kelvin	K
Amount of substance	Mole	Mol
Luminous intensity	Candela	cd

---

[10] https://en.wikipedia.org/wiki/Ounce

[11] https://en.wikipedia.org/wiki/Avoirdupois_system

Other units are defined from the base units according to SI. For example, area can be expressed as square meters, i.e., m², and speed can be expressed as meters per second, i.e., m s^{-1}.

The USA generally does not use the SI, but rather uses the imperial or British system of units. The remainder of this appendix discusses units relevant for energy that are commonly used in the USA.

Length
Various measures are used for length, including inches, feet, yards, and miles. One foot is 12 inches; one yard is 3 feet; and one mile is 1760 yards or 5280 feet.

Area
Area is the size of an enclosed region.[12] For example the area of a rectangle is the product of its width and its breadth, so a rectangle with sides 8 ft and 3 ft has an area of 24 ft², sometimes written as 24 ft². A circular area with radius r has an area of π r², where π (pronounced "pi") is about 3.14. Thus a circle with a radius of 3 ft has an area of about 28.3 square feet, or 28.3 ft².

Volume
Volume is the space occupied by a three-dimensional object. The volume of a cuboid such as shown in Fig. 2.14 is height × length × depth. A cube is a special case of a cuboid for which the height, length, and depth are all equal. So the volume of a cube with sides of 1 ft is $1 \times 1 \times 1 = 1$ cubic feet, often written as 1 ft³ or 1 CF. Amounts of gas are often stated as MCF for 1000 ft³, MMCF for 1,000,000 ft³, BCF for a billion cubic feet (1,000,000,000 or 10⁹ ft³), or TCF for a trillion cubic feet (10¹² ft³). See the density discussion below for a better understanding of volumes of gas.

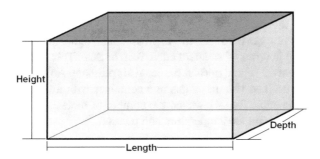

Fig. 2.14 Cuboid

Other imperial units are used for volume also. For example, 1 US gallon is about 0.1337 ft³. Oil volumes are often stated in barrels. An oil barrel is 42 US gallons or about 5.615 ft³. The symbols, b, B, and bbl are used to denote a barrel of oil. Mb, MB, or Mbbl denote 1000 barrels. MMb, MMB, or MMbbl denote 1,000,000 barrels.

---

[12] The region may or may not be contiguous.

Rates

Flow rates are measured in either volumetric or mass terms. Common flow rates for oil are barrels per day (BPD), MBPD (thousand barrels per day), and MMBPD (million barrels per day). Sometimes rates are expressed on an annual basis, in which case it is common to not state "per year." Common rates for gas are MCFD (thousand cubic feet per day), MMCFD (million cubic feet per day), or BCFD (billion cubic feet per day). Although not explicitly written, this almost always refers to standard cubic feet,[13] and is sometimes written as MSCFD, MSCFD, MMSCFD, or BSCFD. Sometimes gas rates are expressed in terms of BTUs, such as MMBTU/D, i.e., million BTU per day.

Flow rates can also be written in terms of mass, such as tons per day. This is more common internationally than in the USA.

Weight or Mass

Weight, also known as mass, is measured in ounces, pounds, or tons. As mentioned above, there are two common uses of the word ton. In common usage, ton typically refers to a short ton, i.e., 2000 pounds. However, it may refer to a long ton, i.e., 2240 pounds.

Density

Density is defined as the mass per unit of volume. The density depends on pressure and temperature. The density of water at atmospheric temperature and pressure is about 62 pounds per cubic foot. The density of oil depends not only on the temperature and pressure, but also on its composition. The density of crude oil at atmospheric conditions is typically between 54 and 65 pounds per cubic foot. When its density is less than that of water, it floats on top of water. See the main text for a definition of API gravity which is commonly used to specify the density of crude oil.

The density of any gas is very sensitive to changes in pressure and temperature. For example, if the pressure doubles then the density approximately doubles.[14] Since the density of a gas is so sensitive to pressure and temperature, amounts of gas are often reported in terms of standard cubic feet, or SCF. This is the amount of gas in one cubic foot at 60 °F and normal sea-level air pressure. An SCF is not strictly a measure of volume. The volume of gas in a container with a volume of one cubic foot is always one cubic foot. However, the number of moles[15] of gas in that cubic foot varies depending on the temperature and pressure.

---

[13] Standard conditions specify both pressure and temperature. Different countries use different pressures and temperatures to denote standard conditions. One standard that is used by the Society of Petroleum Engineers and the US Government agency OSHA is 14.696 pounds per square inch and 60 °F. It is important to specify the pressure and temperature for any gas since its physical properties are very sensitive to both pressure and temperature.

[14] The precise relationship between density, pressure, and temperature is heavily studied. It is often modeled with a so-called equation of state.

[15] A mole of gas contains about 6×10^{23} molecules.

It is often convenient to know whether something is lighter or heavier than air or water. Specific gravity is a measure of this. The specific gravity of a liquid is its density divided by the density of water. So if a liquid, like crude oil, has a density of 60 pounds per cubic foot, then its specific gravity is 60/62 = 0.97. Since this is less than 1, that oil will float on water. Similarly the specific gravity of a gas is the density of that gas divided by the density of air. Natural gas is mostly methane which is less dense than air when at the same temperature and pressure as the air. So its specific gravity at atmospheric conditions is less than 1, and it will rise upwards in air. However, there are concerns that if a cargo of liquefied natural gas (LNG, which is mostly methane) leaked then its density will be greater than that of air since it will be cold. If so, it would settle below air at ground or sea level and create a risk of fire or explosion.

Temperature
In the USA temperature is measured in degrees Fahrenheit, written as °F. At atmospheric pressure water freezes at 32 °F and boils at 212 °F. In the metric system, temperature is measured in degrees Celsius. At atmospheric pressure water freezes at 0 °C and boils at 100 °C.

Pressure
Pressure is a measure of the force exerted per unit of area. If you have ever dived to the bottom of a swimming pool, or gone scuba diving, then you have probably felt your ears respond to the change in pressure. The pressure in oil and gas reservoirs that are below the surface of the earth can be quite high. Pressure is measured in various units, with pounds per square inch being common in the USA. The atmosphere exerts a pressure of about 14.7 pounds per square inch at sea level. Sometimes pressure is stated in "atmospheres," which is a multiple of atmospheric pressure. So 300 pounds per square inch is about 20 atmospheres. Absolute pressure is the actual pressure. It is often denoted as psia. Gauge pressure is the pressure reported on pressure gauges. It is the pressure in excess of atmospheric pressure, and is denoted as psig. The pressure at sea level is about 0 psig. 300 pounds per square inch is 300 psia and about 285 psig.

Energy
Several units are used to measure energy. One common unit is a British thermal unit (BTU or Btu). A common unit for energy in the SI system is a joule (J). One BTU is the amount of energy required to raise the temperature of one pound of water by one degree Fahrenheit. One BTU is about the amount of energy produced by burning a single wooden kitchen match.[16] Another measure is a calorie, which is the amount of energy required to raise the temperature of one gram of water by one degree Celsius. Prefixes are used for BTUs. One MBTU is a thousand BTUs and one MMBTU is a million BTUs.

---

[16] https://en.wikipedia.org/wiki/British_thermal_unit

One therm is 100,000 BTUs. A decatherm is 10 therms, or 1 MMBTU. A quad is 10^{15} BTUs.

Another measure of energy is a kilowatt-hour (kWh). This is the amount of energy provided by an energy source with a power of 1 kW (defined below) during 1 hour. 1 kWh is about 3412 BTUs.

Energy can also be used to do work. One BTU is enough energy to raise one pound by 778 ft.[17]

Power

Power is the rate of energy use or of energy delivery per unit of time. For example, the handheld blow torch shown in Fig. 2.15 is rated at 2500 BTUs per hour. This is the rate at which it delivers heat, i.e., its maximum power. The blue fuel cylinder contains 14.1 ounces of propane, whose heat content (energy) is about 19,000 BTUs. If the valve on the torch is fully open it will use up that energy at a rate of 2500 BTUs per hour, and be fully depleted in about 19,000/2500 = 6.4 hours, although the actual time will depend on various factors such as the amount of propane that is used during ignition. If the valve is less than fully open, then the power delivery will be less than 2500 BTUs per hour.

Fig. 2.15 Handheld blow torch

The power output of the blow torch must not be confused with the temperature of the flame. The temperature of the flame depends not only on the power of the

---

[17] https://en.wikipedia.org/wiki/British_thermal_unit

blow torch, but also on the type of fuel and details about the combustion process, which are fascinating but way beyond the scope of this book.

BTUs per hour is not a very common measure of power. More common measures are watts, kilowatts (kW, 1000 watts), megawatts (MW, 1,000,000 watts), and horsepower. A watt (W) is an SI unit equal to 1 J/s. The power rating of some light bulbs is 100 W, i.e., 0.1 kW. The product of the power and time is a measure of energy; for example 1 kWh is the amount of energy used or generated during 1 hour by a power device or source of 1 kW. For example, if a 100 W bulb is on for 2 hours then it will consume 2 hours × 0.1 kW = 0.2 kWh.

A common measure of power in engines is horsepower. Steam engines became common during the Industrial Revolution. James Watt played a central role in the development and marketing of steam engines. He is associated with the horsepower measure. He invented that unit in order to compare the power output of a steam engine with that of a horse. One horsepower is about 746 watts.

Refrigeration power is sometimes expressed in terms of tons, which is not to be confused with a measure of weight! One ton of cooling is the rate of heat transfer required to freeze one short ton of water into ice in 24 hours.[18]

It is very important not to confuse energy and power!

Combining Different Types of Units
It is often convenient to combine quantities of different types of units. For example, a gas well may produce 3 MMCFD gas and 100 BPD of oil (or condensate). One method for combining such rates is known as energy equivalence. 6 MCF of gas has about the same amount of energy as 1 barrel of oil, so 6 MCF of gas has 1 barrel of oil equivalent, written as 1 BOE. Thus the example gas well produces 3000 MCFD/6 MCF per BOE + 100 BPD = 600 BOE per day. This can also be written as 3000 MCFD +100 BPD × 6 MCF per BOE = 3600 MCFGE per day or 3.6 MMCFGE/D.

Another combination method is to use tons of oil equivalent (TOE). Some conversion factors are shown in Table 2.12.

Table 2.12 Tons of oil equivalent (TOE) conversion factors	Commodity	Amount	TOE
	Natural gas	1 BCF	0.024
	Natural gas	1 trillion BTU	0.025
	LNG	1 million tons	1.169
	Oil	1 barrel	0.146
	Hard coal	1.5 tons	1
	Lignite or sub-bituminous coal	3 tons	1
	Electricity	12 MWh	1
	Ethanol	1 ton	0.68

---

[18] https://en.wikipedia.org/wiki/British_thermal_unit

Unit Conversion

You will often need to convert between different units. There is a foolproof method for doing this correctly, which involves writing the units explicitly. For this to work, you need to recognize that the word "per" can be interpreted as division. So "miles per hour" is miles divided by hours. As a conversion example, suppose you want to know the length of 18 inches in feet. Since there are 12 inches in a foot, the calculation can proceed as follows:

$$\frac{18\,\text{in.}}{12\,\text{in.}/\text{ft}} = \frac{18\,\text{in.}}{12\,\text{in.}/\text{ft}} = \frac{18\,\text{in.} \times \text{ft}}{12\,\text{in.}} = 1.5\,\text{ft}\,.$$

The number 1.5 comes from dividing 18 by 12. The "foot" unit moves from the bottom of the fraction to the top since $\frac{1}{1/x} = x$ for any x. The "inches" unit cancels between the bottom and top, leaving only the "foot" or "feet" unit.

As another example, 640 acres is 1 square mile or 1 mile^2, and 1 mile is 5280 ft. Suppose you want to know the area of 640 acres in square feet:

$$\frac{640\,\text{acres}}{640\,\text{acres per mile}^2} = 1\,\text{mile}^2$$

$$\frac{1\,\text{mile}^2}{\left(1\,\text{mile}/5280\,\text{ft}\right)^2} = \frac{1\,\text{mile}^2}{\left(1\,\text{mile}\right)^2/\left(5280\,\text{ft}\right)^2} = \left(5280\,\text{ft}\right)^2 = 27{,}878{,}400\,\text{ft}^2$$

English/Metric Conversion

Sometimes you will need to convert between English and metric units. Table 2.13 shows some conversion factors.

Table 2.13 English/metric conversion factors

	English	Metric equivalent
Length	1 inch	2.54 cm
	1 foot	0.3048 m
Area	1 square inch = 1 in^2	6.452 cm^2
	1 square foot = 1 ft^2	0.0929 m^2
	1 acre	0.4047 ha (hectare)
Volume	1 cubic foot = 1 ft^3	0.0283 m^3
Weight	1 pound = 1 lb.	0.454 kg = 454 g
	1 short ton	907 kg = 0.907 metric ton = 0.907 tonne
Density	1 lb./ft^3	16.02 kg/m^3
	1 lb./US gal	119.8 kg/m^3
Pressure	1 psi (pound per square inch)	6895 Pa
	1 atmosphere	101,325 Pa

For example, 2000 psi = 2000 psi × (6895 Pa/psi) = 13,790,000 Pa = 13.79 MPa (megapascals).

Temperature

Temp in $^\circ$C = (Temp in $^\circ$F – 32)/1.8
Temp in $^\circ$F = 1.8 × (Temp in $^\circ$C) + 32

Energy

Various conversion factors are shown in Table 2.14. As examples, 1 calorie is 4.184 J and 1 BTU is 2.93 10^{-4} kWh.

Table 2.14   Energy conversion table

	Equivalent				
	J	Cal	BTU	kWh	TOE
J	1	0.239	0.000948	2.78 10^{-7}	2.37 10^{-11}
Cal	4.184	1	0.00397	1.16 10^{-6}	9.92 10^{-11}
BTU	1055	252	1	2.93 10^{-4}	2.50 10^{-8}
kWh	3.60 10^6	8.60 10^5	3412	1	8.53 10^{-5}
TOE	4.22 10^{10}	1.01 10^{10}	4.00 10^7	1.17 10^4	1

Power

One watt is 0.00134 horsepower, and one horsepower is 746 watts.

Sources

Figures

Number	Source
2.1	https://electrek.co/2018/09/24/tesla-powerpack- battery-australia-cost-revenue/
2.9	https://en.wikipedia.org/wiki/Coal#/media/
2.10	https://commons.wikimedia.org/wiki/
2.13	Photo by David Johnson, Wikipedia

Tables

Number	Source
2.5	http://www.roperld.com/science/TeslaModelS.htm
2.9	https://en.wikipedia.org/wiki/Heat_of_combustion
2.10	https://www.eia.gov/environment/emissions/co2_vol_2.mass.php
2.11	https://en.wikipedia.org/wiki/International_ t11.10 System_of_Units
2.12	BP Statistical Review of World Energy, 2019

Energy Use

3

3.1   World Energy Consumption by Fuel Type

Consumers access energy in many different forms including gasoline, diesel, jet fuel, natural gas, and electricity. Gasoline, diesel, and jet fuel are all produced by refining crude oil. Some of these liquid fuels can also be produced from coal or agricultural feedstocks instead of oil. For example, biodiesel is made from soybean oil, other plant oils, rendered fats, recycled grease, and algae [1]. It is also possible to convert natural gas into very clean-burning diesel, using a process known as gas to liquids. But currently the quantities of liquid fuels produced with these alternative processes are very small compared to the amount produced from oil.

As explained in Chap. 7, electricity is produced from many different sources including coal, natural gas, nuclear energy, hydropower, solar, wind, and some liquid fuels. Electrical power plants that use coal, natural gas, or liquid hydrocarbon fuels are said to use "fossil fuels." Fossil fuel that is supplied to power plants is referred to as "primary energy" to differentiate it from the energy that is delivered from the plants for end use as electricity. The primary chemical energy in fossil fuels is transformed into electrical energy because electricity is so convenient to use. For example, it is much easier to flick a switch to turn on the lights rather than filling a kerosene lamp and then striking a match to start the wick burning.

Energy economists keep track of the primary energy consumption that is the underlying source of energy. The breakdown of worldwide primary energy consumption in 2019 is shown in Fig. 3.1.

M. Cronshaw, *Energy in Perspective*,
https://doi.org/10.1007/978-3-030-63541-1_3

Fig. 3.1 World energy consumption by fuel type (2019)

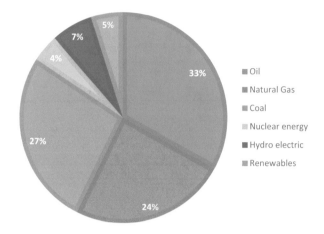

Oil provided the largest amount of primary energy worldwide, although coal was a close second. Natural gas also provided a large amount of primary energy. Between them, these fossil fuels provided 84% of the world's primary energy in 2019. As you drive and see many cars, trucks, and motorcycles on the road, it is easy to understand why oil is such an important primary source of energy, since oil feeds refineries which create gasoline and diesel. In contrast, the widespread use of coal is less apparent to the eye, since almost all coal is consumed by power plants or by industry. But if you visit or live near coal-producing regions such as Wyoming, you are likely to see coal trains with many railcars stretching into the distance; and if you are unlucky enough to encounter such a train at a railroad crossing then you will develop a much better understanding of the huge quantities of coal that provide primary energy.

The use of natural gas is also not always apparent, unless you use natural gas for cooking or have a gas fireplace. However, most of us experience the benefit of energy in the form of natural gas every time we walk into a warm house or take a hot shower, since natural gas is commonly used for residential space and water heating. It is also used as fuel in many electrical power plants.

Later chapters will provide more detail about the use of fossil fuels, and also about the other forms of primary energy shown in Fig. 3.1. Each of these forms can be used to generate electricity. About 7% of worldwide primary energy consumption is in the form of hydroelectricity. Hydroelectricity is regarded as primary energy since it is not created from another energy source. Large dams restrict the flow of rivers in many locations. This can be useful for flood control and also for generating electricity. The water stored behind the dams is released through turbines which generate electricity. Four percent of worldwide primary energy consumption is in the form of nuclear energy that powers nuclear power plants, and 5% of worldwide primary energy is in the form of renewable energy, mostly solar and wind.

Fossil fuels provide about five times as much primary energy as hydropower, nuclear, and renewable energy combined. Fossil fuels are the predominant supply of worldwide primary energy.

The corresponding data for primary energy consumption in the USA are shown in Fig. 3.2. The figure shows that the share of coal in the US primary energy consumption in 2019 was much less than the share in world primary energy consumption. The shares of primary energy from other sources are correspondingly higher.

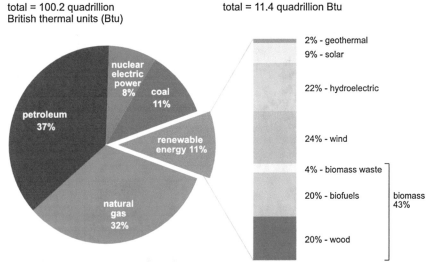

U.S. primary energy consumption by energy source, 2019

Note: Sum of components may not equal 100% because of independent rounding.
Source: U.S. Energy Information Administration, *Monthly Energy Review*, Table 1.3 and 10.1, April 2020, preliminary data

Fig. 3.2 US primary energy consumption

3.2 US Energy Consumption by Sector

Pun intended, it is enlightening to see how primary energy is used. Figure 3.3 shows how energy was used in the USA in 2019. The left side of the figure shows the primary energy types, while the right side shows the various sectors that consumed the energy. The transportation sector includes cars, trucks, buses, planes, trains, and vessels. The industrial sector includes all industrial consumption of energy except for electrical power generation, which is shown separately. The residential sector is for housing, and the commercial sector includes workplaces, government buildings, retail establishments, and hotels.

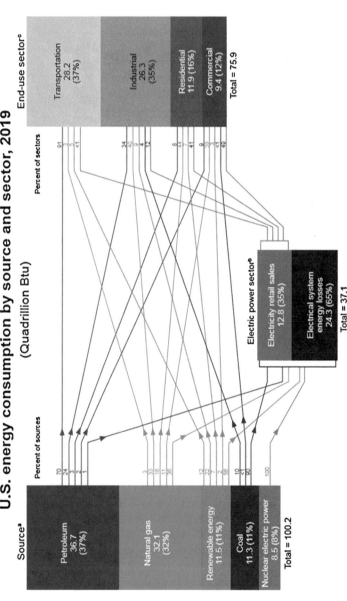

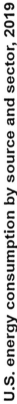

Fig. 3.3   US energy consumption

Fossil fuels provided 80% of primary energy in the USA during 2019, slightly less than the worldwide share of 84%. Correspondingly, renewable energy (including hydropower) and nuclear energy provided slightly more energy in the USA than in the world as a whole. In fact, nuclear energy in the USA provided about twice the share of primary energy compared to the world as a whole. This is indicative of the advanced nature of the US economy compared to the world. The 11% share of the US primary energy in the form of renewables (including hydropower) is also about twice the worldwide share.

The numbers shown above the arrows on the left of Fig. 3.3 indicate how the primary energy is used across the consumption sectors, while the numbers above the arrows on the right side show the relative importance of the types of primary energy in each of the sectors. For example, 91% of the energy used in the transportation sector is provided by petroleum, while only 3% is provided by natural gas, and 5% by renewable energy. Liquid fuels derived from oil, primarily gasoline, diesel, and jet fuel, are ideally suited for use in transportation. Renewable energy in the form of electricity is used to charge electric cars in the USA, while natural gas powers some other vehicles including buses. Natural gas for transportation in the USA is not common, but it is used more in other countries such as Bangladesh.

Almost equal amounts of natural gas are consumed in the US industrial and electric power sectors. The residential and commercial sectors also use natural gas, but in lower amounts. Ninety percent of US coal production is used for power generation. All US nuclear energy is used for power generation.

Notice that only about 35% of the primary energy delivered to the electric power sector is output for electricity sales, while losses account for 65%. There are thermodynamic limits to the efficiency of energy conversion which constrain the energy output of power plants. It may shock you that almost twice as much of the primary energy is wasted in the power sector compared to the electricity that is generated. This is discussed in Chap. 7.

3.3   Trends in Energy Use

The world's consumption of primary energy in 2019 was almost four times as much as in 1965, as shown in Fig. 3.4. This represents an average growth rate of 2.5% per year over the 54-year period. Much of the increase is due to growth in the world population. The world population increased by a factor of 2.3 over the same period, which is an average growth rate of 1.6% per year. The higher growth rate in energy consumption compared to the population growth rate reflects higher energy consumption per person. Energy consumption per person increased at an average rate of 0.9% per year over that period. Economic development is associated with increased use of energy. They go hand in hand. Increased access to energy results in increased economic development, and increased economic activity typically requires additional energy. One could say that increased access to energy has been one of the causes of worldwide economic development.

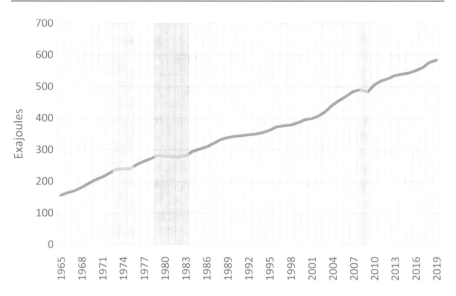

Fig. 3.4 Historical world primary energy consumption

The figure shows that the world's primary energy consumption has not increased smoothly over time. The shaded regions in the figure are periods during which there was a decline or almost no change in primary energy consumption. From 2008 to 2009, primary energy consumption declined by 1.5%. It also declined from 1979 to 1983. From 1973 to 1975 the growth rate averaged only 0.4% per year, much lower than the average over the longer period. World primary energy use seems to be a barometer of world activity. What happened during these unusual periods?

- In October 1973 several oil-producing countries temporarily ceased sending oil to the USA and the Netherlands, since those countries had supported Israel during the Yom Kippur War. Oil deliveries resumed about 6 months later in March 1974. The price of oil quadrupled during that oil embargo, and the USA imposed fuel rationing. Higher prices led to more conservation of energy. This together with the oil export reduction had a significant impact on world energy consumption.

- In 1979 there was a revolution in Iran that resulted in the overthrow of the Shah. The following year, Iraq invaded Iran. Both of these events severely disrupted Iranian oil production. The resulting restriction in oil supply led to a doubling of crude oil prices. The roughly eightfold increase in the oil price from 1973 to 1980 led to significant energy conservation, and a decline in world primary energy consumption from 1979 to 1983.

- In 2007 the world experienced a financial crisis that led to a global economic recession. The investment bank Lehman Brothers failed, and there were debt crises in the USA and Europe. The slowdown in global economic activity led to a decline in worldwide primary energy consumption.

The covid-19 pandemic had profound global impacts in 2020. Many people contracted the disease, and many died. The number of people working from home rose dramatically, and many countries imposed limits on international travel. The changes in commuting and other travel reduced the demand for energy. US oil production dropped by 18% from March to June 2020, and natural gas production dropped by 6%. The number of oil and gas drilling rigs operating in the USA dropped by about 2/3 in response to the drop in demand.

3.4   Energy Use by Region and Country

In the 54 years from 1965 to 2019, economic development has had a huge impact on the global distribution of primary energy consumption, as shown in Fig. 3.5. In 1965 Europe and North America consumed 66% of the world's primary energy. By 2019, their combined share was greatly reduced to 34%. During the same period, the Asian Pacific share of primary energy went up by almost four times, from 12% to 44%. The share shown for Europe has an unusual step up in 1985. Ukraine was added to Europe in this year.

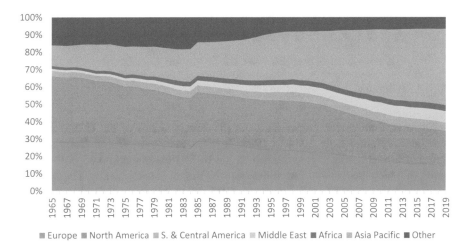

Fig. 3.5 Regional shares of world primary energy consumption

The Asia Pacific region includes China and India, both of which underwent massive economic development during this period. Economic development is associated with increased energy use, both when industrial activity increases and as people use more energy. Figure 3.6 shows that China's share of primary energy changed particularly dramatically during this period, from 3.5% to 24.3%, almost a sevenfold increase. India's share also increased, from 1.4% to 5.8%, more than a four-fold increase. Correspondingly, the US share of primary energy consumption dropped in half, from 33.7% to 16.2%. China's share of primary energy first exceeded the share in the USA in 2009, and the gap between their shares has increased since then.

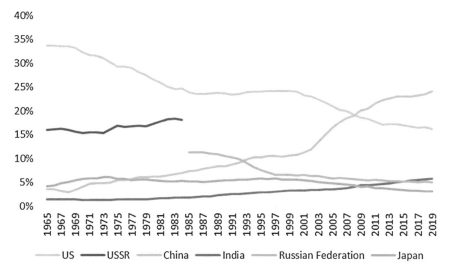

Fig. 3.6 Evolution of primary energy shares

There is a noticeable discontinuity in the energy shares of the USSR and the Russian Federation from 1984 to 1985. In 1991 the Soviet Union (formally the Union of Soviet Socialist Republics, i.e., the USSR) dissolved into 15 post-Soviet states, of which the Russian Federation was the largest geographically. The seeds of this dissolution can be traced back to 1985 when Mikhail Gorbachev became General Secretary of the Politburo and initiated significant liberalization known as perestroika (restructuring), or glasnost (openness). Since the Russian Federation is just one of the 15 post-Soviet states, its share of primary energy is lower than that of the entire USSR. This caused the discontinuity shown in blue lines in the figure.

3.5   Per Capita Energy Use

One measure of energy consumption is the total used within a region or country. It is an important indicator, but large countries will naturally have higher consumption according to this indicator. Another useful indicator is consumption per capita, i.e., per person. Table 3.1 shows per capita annual energy consumption for various

Table 3.1   Per capita annual primary energy consumption (2019)

High per capita consumption		Low per capita consumption	
Country	GJ per person	Country	GJ per person
Qatar	714.3	Egypt	38.7
Iceland	647.8	Colombia	38.2
Singapore	611.6	Peru	35.7
Trinidad and Tobago	511.6	Indonesia	32.9
United Arab Emirates	494.4	Morocco	26.0
Kuwait	389.2	India	24.9
Canada	379.9	Philippines	18.7
Norway	328.5	Sri Lanka	16.8
Saudi Arabia	322.0	Pakistan	16.4
Oman	304.3	Bangladesh	10.8

countries in 2019. The table shows a very wide range of energy consumption per person, from about 11 GJ per person in Bangladesh to 714 GJ per person in Qatar. It is hard to imagine people in Qatar using 65 times as much energy as people in Bangladesh, even though their levels of economic development are quite different. People can only consume so much energy for heating, cooling, transportation, and so on. So there must be some other explanation for the wide variation in per capita energy consumption. Geography cannot be the answer since even though Qatar, the United Arab Emirates (UAE), and Kuwait are in similar locations, Qatar's per capita consumption is much higher than that in the UAE and Kuwait. Industrial activity provides the answer.

If an energy source like natural gas is used as an input to an industrial facility such as a chemical plant, then that energy is treated as having been consumed within the country. The population has not consumed the energy for personal use, but the energy is still counted as having been consumed. Qatar and Trinidad and Tobago each have large reserves of natural gas. They monetize these reserves partly by liquefying the gas for export (see Chap. 5) and partly by using the gas as feedstock to make products such as ammonia, urea, and methanol (see Chap. 8). Gas that is liquefied is not counted as having been consumed. However, gas that is used as feedstock for chemical manufacturing is regarded as having been consumed. Ammonia and urea are important for use as agricultural fertilizer. Methanol is an important feedstock for other chemical plants that make acetic acid and formaldehyde, which in turn are used in products like adhesives, foams, plywood subfloors, solvents, and windshield washer fluid. Methanol can also be used as a fuel additive or to make other fuel additives. An appendix to this chapter provides some details about the top four countries in Table 3.1. Briefly, there are three different reasons for the high per capita consumption. Qatar and Trinidad and Tobago have high consumption since they each uses natural gas to make petrochemicals. Iceland has an abundant supply of hydroelectric power, some of which they use for aluminum smelting, which is an energy-intensive process. Singapore provides large amounts of bunker fuel for shipping and aviation. Each of these activities (chemicals, smelting, and bunkering) uses large amounts of energy, although the feedstock for the chemical plants serves dual roles: both as chemical inputs for chemical products and as energy for the transformation process from inputs to outputs. This is discussed more in Chap. 8.

3.6   Cross-Country Energy Supply Comparisons

The energy supply mix is quite different in various countries. This is illustrated by the listing of countries shown in Table 3.2.

Table 3.2   Countries significantly reliant on a specific energy type (2019)

Fossil fuels			Non-fossil fuels		
Oil	**Natural gas**	**Coal**	**Nuclear**	**Hydro**	**Renewables**
Singapore	Trinidad & Tobago	South Africa	France	Norway	Portugal
Hong Kong	Uzbekistan	China	Sweden	Iceland	Brazil
Sri Lanka	Turkmenistan	India	Ukraine	Ecuador	Finland
Iraq	Qatar	Kazakhstan	Finland	Brazil	Germany
Ecuador	Bangladesh	Vietnam	Switzerland	Switzerland	Sweden
Saudi Arabia	Belarus	Poland	Czech Republic	Sweden	United Kingdom
Morocco	Iran	Indonesia	Hungary	Venezuela	Spain
Greece	Azerbaijan	Philippines	Belgium	New Zealand	Chile
Belgium	Algeria	Czech Republic	South Korea	Peru	New Zealand
Portugal	Oman	Taiwan	Spain	Austria	Italy

Energy share exceeds 60% in the shaded cells
Some countries may be missing from the table since they are included in aggregated "other" regions in the original source data

The table shows three types of fossil fuel energy and three types of nonfossil fuel energy. For each type of energy, the countries are listed in decreasing order of the share of that type of energy which is consumed within the country. For example, Singapore was the country for which oil provided the largest share of primary energy and Hong Kong was the location in which oil provided the second largest share. Portugal was the country for which renewables (excluding hydro) provided the largest share, and Brazil was the country for which renewables provided the second largest share. Some countries appear twice in the table. As examples, the Czech Republic used significant shares of both coal and nuclear energy, Belgium used significant shares of oil and nuclear energy, Ecuador used significant shares of both oil and hydro, and Sweden used significant shares of hydro and renewables.

Countries that do not appear in the table, such as the USA, consumed a relatively diverse mix of energy, so no single source of energy was a dominant supply of its energy. The shaded cells in the table represent the other extreme. A cell is shaded if that type of energy supplied at least 60% of the energy consumed in that country. For example, Singapore had a very nondiversified energy mix. 86% of the energy consumed in Singapore was in the form of oil, and 89% of the energy consumed in Trinidad and Tobago was in the form of natural gas.

It is striking that nine countries relied on natural gas for more than 60% of their energy consumption. All of these countries, except Belarus, have large reserves of natural gas, so it is natural for them to use this indigenous resource to serve their domestic energy needs. Belarus is exceptional since it does not have substantial reserves of natural gas. However, it is a close ally of Russia, which has built pipe-lines to deliver gas to Europe. Those pipelines traverse Belarus, enabling delivery of gas in Belarus. It is common for countries that allow pipelines to cross their territory to receive natural gas in a transaction known as a transit fee.

South Africa receives more than 60% of its energy needs from coal. It has large supplies of coal, so it is natural to use those supplies for its energy needs. The situ-ation with oil is more varied. Ecuador and Saudi Arabia each have large oil reserves, and use significant quantities of oil domestically. Singapore has no hydrocarbon reserves,[1] and imports all its crude oil and natural gas. However it is a center for trading and refining in Asia, and it provides large amounts of fuel for ships and airplanes. It is the fifth largest exporter of refined oil products in the world. Three large oil refineries are located there. In 2019, Singapore imported 49.6 million tons of crude oil and 112.4 million tons of petroleum products [1]. It exported 86.1 mil-lion tons of petroleum products.

Only one country uses hydropower for more than 60% of its primary energy consumption: Norway. As of 2018 the small country of Norway, with about 5.4 mil-lion inhabitants, had 1121 hydropower stations.[2] Hydroelectricity is generated mainly by damming water supplies, and then releasing the water through turbines (see Sect. 7.2.5). It requires two key things: adequate water supply (i.e., rain or snow) and sufficient change in geographic elevations so that the potential energy of stored water can be released by falling to a lower elevation.[3] Norway has both water and elevation changes (mountains). Due to its abundant supply of hydropower, Norway hosts industries that consume large amounts of electricity, including the production of aluminum, industrial chemicals, and ferrosilicon.[4]

3.7   Energy Use by Sector

3.7.1   Residential

The main use of energy in the residential sector is for heating and cooling. Figure 3.7 shows that heating and cooling together used about half of the energy consumed in this sector in the USA in 2015. Space heating used about 27%, space cooling (i.e., air-conditioning) used about 12%, and water heating (for showers, baths, laundry, and dishwashers) used about 13%. Lighting used about 7% of the energy, and appli-ances (fridges, freezers, washers, dryers, dishwashers, television, and cooking appliances) together used about 18%. Computers in homes used less than 2% of the energy consumed in the residential sector.

---

[1] https://www.eia.gov/beta/international/country.cfm?iso=SGP

[2] Statistics Norway, https://www.ssb.no/en/statbank/table/10431/tableViewLayout1/, retrieved June 29, 2020.

[3] See Sect. 7.2.4 for a discussion.

[4] https://www.ssb.no/en/elektrisitet

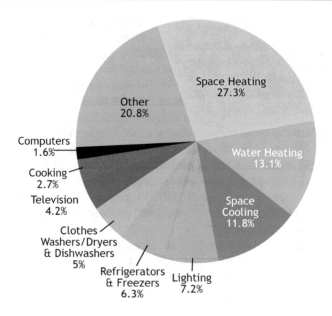

Fig. 3.7 US residential energy usage (2015)

It is interesting to consider the different types of energy that are used to provide energy services to the residential sector. The "percent of sectors" numbers in Fig. 3.3 provide data. Natural gas and electricity provided almost equal amounts of energy: 44% and 41%, respectively. Natural gas is an excellent and convenient fuel for space heating and for water heating. Petroleum and renewables also provided about equal amounts, but much lower than natural gas or electricity: 8% and 7%, respectively. Fuel oil is used for heating some homes in the USA. You have doubtless seen rooftop solar installations in homes. This is an example of renewable energy used in the residential sector. Also, wind turbines provide energy to some commercial and residential buildings.

3.7.2 Commercial

The commercial sector includes offices, hospitals, schools, police stations, places of worship, warehouses, hotels, and shopping malls. Figure 3.8 shows the breakdown of energy use in the US commercial sector in 2012.

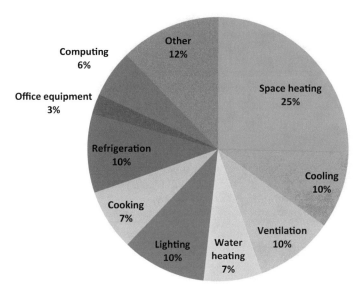

Fig. 3.8 US commercial energy use

Space heating consumed the largest share of energy in the commercial sector, 25%, which is similar to the share for space heating in the residential sector. Cooling, ventilation, refrigeration, and lighting each consumed 10% of the energy used. Water heating, cooking, and computing each consumed slightly smaller shares.

Electricity and natural gas provided 49% and 39% of the energy for this sector in 2019 (see Fig. 3.3). Petroleum and renewables supplied much smaller shares: 9% and 3%, respectively. Fuel oil is used for heating in some US markets. The renewable share is smaller than in the residential sector, probably due to the smaller amount of roof space for solar in multistory buildings compared to residences.

3.7.3 Transportation

Table 3.3 shows that cars, trucks, vans, SUVs, and motorcycles consumed slightly more than half of all energy used in the US transportation sector in 2018. Commercial and freight trucks consumed about a quarter of the energy in the sector, and aircraft consumed 9%.

It may seem curious that pipelines are included in the transportation sector. However, significant amounts of energy are required for the pumping and compression that are used to move liquids and gases in pipelines.

Table 3.3 Transportation energy use in the USA (2018)

Mode	Percentage of transportation energy
Light-duty vehicles (cars, small trucks, vans, SUVs, and motorcycles)	55%
Commercial and freight trucks	24%
Jets, planes, and other aircraft	9%
Boats, ships, and other watercraft	4%
Trains and buses	3%
Military	2%
Pipelines	3%
Lubricants	<1%

Fuel tanks in gasoline- or diesel-fueled cars have a capacity of 12–20 gallons. The capacities of other types of transportation vary widely:

	Thousand gallons
Petroleum rail car	30
Boeing 747	50 to 60
Bulk grain carrier vessel	400 to 800
Large cruise ship	1000 to 2000
Panamax container ship	1500 to 2000

Source: https://response.restoration.noaa.gov/about/media/how-much-oil-ship.html

Figure 3.3 shows that petroleum provides almost all of the energy used in the transportation sector. Gasoline, diesel, and jet fuel have high energy densities (BTU per pound) which make them well suited for transportation. Natural gas provides a small share (3%); renewable energy provides another small share (5%). This is the share of energy supplied for transportation directly by renewable energy. However, the figure also shows that electricity generated from other sources of energy is also used for charging the batteries in electric vehicles. Some of that electrical energy comes from renewables as shown in the figure.

Natural gas is used as a vehicle fuel in several countries. It is stored as compressed natural gas (CNG) in high-pressure cylinders that serve as fuel tanks. About 15% of automobiles in Argentina ran on CNG in 2008.
Source: https://archive.is/20081120221031/http://www.ngvglobal.com/en/country-reports/latin-america-ngvs-an-update-report-02074.html

3.7.4   Electricity

Electricity is an extremely convenient source of energy. It is almost always available at the flick of a switch for customers who are connected to an electricity grid, except during the rare occurrences of blackouts. Chapter 7 provides information about the variety of ways in which electricity is generated, and also about how it is supplied to customers.

Figure 3.3 shows the mixture of energy sources that is used for US electricity generation. 90% of US coal was used for electricity generation and 100% of nuclear in 2019. This was supplemented by 36% of natural gas and 56% of renewable energy. It is almost certain that the use of renewables will increase over time due to consumer interest and the fact that no carbon dioxide is emitted during the generation of electricity by solar and wind.

3.7.5 Industrial

Figure 3.9 shows the use of energy for all purposes (fuel and nonfuel) in various parts of the industrial sector in the USA in 2014. The chemicals subsector used the most. This industry receives natural gas, which is primarily methane, ethane, propane, and butane, and converts them to other products. Energy is used for this conversion, which explains some of its large share. However, many of the items produced by the chemical industry are not valued for their energy content, but rather for other properties. For example, ethane is an excellent fuel, but it is also a feedstock for the manufacture of ethylene which is a precursor for polyethylene. You may not be familiar with polyethylene, but you have probably used it regularly. It is used to make plastic bags and other plastic containers.[5] As another example, propane is an excellent fuel and is used as such, but it is also a feedstock for the manufacture of propylene which is a precursor for polypropylene that is used for containers, furniture, and clothing.

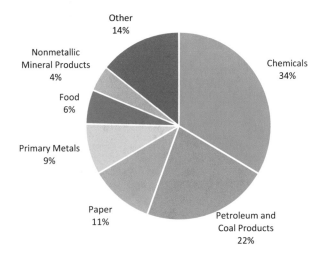

Fig. 3.9 US industrial energy use by industrial subsector (2014)

---

[5]You may see the acronyms LDPE and HDPE on some containers. These stand for low-density polyethylene and high-density polyethylene, respectively. This is discussed further in Chap. 8.

Petroleum and coal products used the second largest amount of energy in the industrial sector. Most of the energy used in the petroleum sector was fed to petroleum refineries. They receive crude oil as an input and produce a variety of products such as gasoline, diesel, and jet fuel (see Sect. 4.6 for a description). Refineries do consume some energy for heat and pumping, but most of the energy in the crude oil input is not consumed in a refinery. Instead refineries transform the crude oil into saleable products that contain large amounts of energy.

Paper and paperboard manufacture also requires significant amounts of energy. The manufacturing process involves substantial amounts of heat to dry pulp that is a combination of fibrous material, such as wood and water.

Figure 3.10 shows that natural gas and petroleum are the primary sources of energy for the industrial sector, with shares of 40% and 34%, respectively. As mentioned above, crude oil is a feedstock for refineries, and natural gas is a feedstock for the chemical industry. It may seem surprising that the industrial sector uses biomass for more than 9% of its energy. About 60% of this biomass is wood and wood-derived fuel. Wood can be used to generate heat and electricity.[6] 33% of the biomass is classified as losses and coproducts from the production of ethanol and biodiesel.[7] Most of the gasoline sold in the USA contains some ethanol. It is manufactured from corn. Coal supplies a small share of the energy used by industry; the manufacture of some metals uses coal.

Fig. 3.10 Energy used in the US industrial sector (2019)

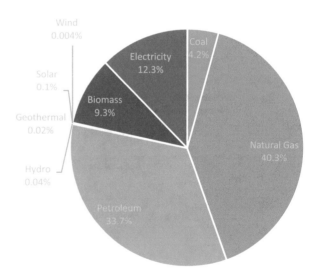

---

[6] US EIA Table 10.2b June 2020.
[7] Ibid.

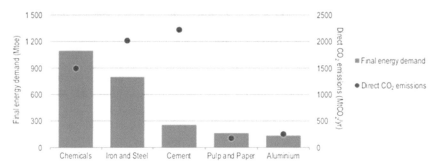

Notes: *Final energy demand* for chemicals includes feedstock, and, for iron and steel, it includes energy use in blast furnaces and coke ovens. *Direct CO_2 emissions* includes energy and process emissions in the industry sector. Mtoe = million tonnes of oil-equivalent.

Fig. 3.11 Global energy demand and direct CO_2 emissions by industry sector (2017)

Figure 3.11 shows global final energy demand and carbon dioxide emission for various industrial sectors. The final energy demand includes both energy used for processing and energy in the feedstock. The chemical sector has the highest final energy demand primarily since hydrocarbons are used as feedstocks for the manufacture of the chemicals (see Chap. 8 for details). The iron and steel sector uses a great deal of energy for processing. The energy consumption in the other sectors in the figure is substantially lower. Interestingly the cement sector had the highest emissions of carbon dioxide. This is a consequence of the chemical process in the manufacture of cement. A key process in the manufacture of cement is "calcining" during which calcium carbonate (limestone with chemical formula $CaCO_3$) is transformed into calcium oxide (lime, CaO). Calcining results in carbon dioxide being released as a by-product.

3.8 Summary

This chapter describes the various types of energy that are used worldwide and in the USA. It also provides a breakdown of how energy is used in the residential, commercial, transportation, and industrial sectors, and shows how energy use has grown faster than the world population. Increased energy use is associated with the global economic development that has occurred in the last several decades.

As would be expected, larger countries tend to consume more energy than smaller ones. But energy consumption per capita is high in several small countries, including Qatar, Iceland, Trinidad and Tobago, and Singapore. The high per capita consumption is explained by industrial activity. Both Qatar and Trinidad and Tobago consume large amounts of natural gas to produce petrochemicals. Iceland uses its abundant supply of hydropower for the energy-intensive process of aluminum smelting. Singapore does not have significant indigenous energy resources, but it has a large refining and chemical sector and provides large amounts of fuel for ships and airplanes.

The sources of energy vary widely in different countries. Some countries rely heavily on oil, others on natural gas, and others on hydroelectricity. For the most part, energy supplies are dictated by indigenous resources, with Singapore being a notable exception.

The chapter also provides an overview of how energy is used in various sectors. The transportation sector relies heavily on petroleum since liquid fuels are an excellent store of energy. The residential and commercial sectors use energy for space and water heating, cooling, and lighting. The residential sector also uses energy for appliances such as washing machines, driers, dishwashers, ovens, and cooktops.

3.9   Questions

1. The energy released by a match is about 1 BTU. Figure 3.3 shows that the USA consumed 100.2 quads in 2019. How many matches would yield the same amount of energy?
2. Figure 3.3 shows that the USA consumed 28.2 quads for transportation in 2019. Table 3.3 shows that about 55% of US transportation energy was for light-duty vehicles in 2018. Assume that the same percentage was used in 2019, and that the average fuel usage for light-duty vehicles was 25 miles per gallon.

   (a) Estimate the total number of miles driven by US light-duty vehicles in 2019. (Hints: Neglect diesel-fueled vehicles, and assume that the heat content of gasoline is 124 MBTU per US gallon.)
   (b) There are about 193 million light-duty vehicles in the USA. Estimate the average miles traveled per year per vehicle.

3. A Boeing 747 aircraft burns about 5 gallons of fuel per mile. A fully loaded 747 carries about 500 passengers.

   (a) How many passenger-miles does a 747 achieve per gallon?
   (b) If a car has fuel economy of 25 miles per gallon and carries two people how many passenger-miles does it achieve per gallon?

4. The "percentage of sources" numbers on the left side of Fig. 3.3 show what percentage of each primary energy type is consumed by each of the end-use sectors. The "percent of sectors" numbers on the right side of the figure show the percentages of each type of primary energy used by each end-use sector. Use the data in the figure to confirm the "percent of sectors" numbers for the industrial end-use sector. (The numbers will not be exact due to rounding.)
5. The 2020 BP Statistical Review of World Energy reports US consumption quantities in 2019 as shown in the table below.

   (a) How many barrels of oil were consumed in 2019?
   (b) How many BCF of gas were consumed in 2019?
   (c) Complete the missing entries in the table.
   (d) What conversion factor for oil was used to calculate the number of *joules* per barrel?

(e) What conversion factor for gas was used to calculate the number of *joules* per MCF?

	Oil	Natural gas	Coal	Nuclear	Hydro	Renewables	Total
Quantity	19,400	81.92					
Units	MBD	BCFD					
Exajoules	36.99		11.34	7.60	2.42	5.83	94.65
% of total	39.1%						100.0%

6. Figure 3.7 shows that lighting consumed 7.2% of the energy consumed in the US residential sector in 2015. An LED light bulb consumes about 18% of the power used by an incandescent bulb with the same brightness. Show that residential energy use in the USA would have dropped by about 5.9% if all households had replaced all incandescent bulbs with LED bulbs.

7. A 1976 vintage refrigerator/freezer in the USA consumed about 1800 kWh per year, while a 2001 vintage unit consumed about 600 kWh per year.[8] So energy consumption in modern fridges is about 1/3 as high as that in old fridges. If this energy savings had not occurred, how much higher would US residential energy use have been in 2015 (see Fig. 3.7).

8. Figure 3.3 shows that 37% of US energy consumption in 2019 was used for transportation. Suppose that the government had mandated improvements in fuel economy for all vehicles that improved average fuel economy (miles per gallon) by 10%. What would have been the reduction in US energy consumption? Assume that the improved fuel efficiency would not affect the percentage of energy used for transportation.

9. [Requires Excel] Look up gross domestic product (GDP, a measure of economic activity) per capita on the Internet for some of the countries listed in Table 3.1. (Any year of GDP data is fine.) Plot GDP per capita against energy consumption per capita. What do you notice?

10. Pump stations 3 and 4 on the Trans-Alaska Pipeline System have electric pumps to move oil through the pipeline. Turbine generators rated at 12.9 MW are used to supply power to the pumps. How much energy in BTUs does a generator provide per day if it runs at full load?

11. The appendix notes that Singapore provided about 2.1 EJ of fuel for ships (marine bunkering). Ships use heavy residual oil for fuel. Singapore reports annual bunker sales of about 50 million tons annually.

(a) What is the heat content of the bunker fuel in GJ per ton?
(b) Large ocean-going vessels can carry about two million US gallons of fuel. The density of residual fuel oil is about 8.3 pounds per gallon. What is the weight of this fuel in tons?
(c) How many large vessel loads of fuel are in the annual bunker sales of 50 million tons?
(d) The port of Singapore reports about 140,000 vessel arrivals per year of all types (container ships, freighters, bulk carriers, tankers, passenger vessels, barges, and tugs). Why is the answer to part (c) so much lower than the total number of arrivals.

---

[8] https://www.aceee.org/files/proceedings/2004/data/papers/SS04_Panel11_Paper02.pdf

Appendix: Country Analyses

This appendix provides some analyses of energy consumption and production in countries that had high per capita energy consumption in 2019 (see Table 3.1). Those who compile energy data must decide how to report energy that is used for industrial purposes. It appears that the standard approach for such energy flows is to treat them as consumption. The flow is not used to provide energy services to the inhabitants, such as heating, cooling, or transportation. However, the energy is used up in the country, so it seems reasonable to refer to it as consumption. This gives rise to the question of why per capita consumption is so high in several countries. The four subsections in this appendix provide answers for Qatar, Iceland, Singapore, and Trinidad and Tobago. The locations and climates in these countries are so different that the answer cannot be due to geographical phenomena affecting the demand for heating, cooling, or desalination.

Qatar

The country with the highest per capita energy consumption in Table 3.1 is Qatar, which is located in the Middle East. Qatar has huge reserves of natural gas (see Table 5.2). Its population in 2019 was small, less than three million people. Hence it relies on exports in order to monetize those large gas reserves. Fig. 3.12 shows the amounts of gas that were produced and consumed in Qatar, as well as the net production, i.e., production minus consumption. The net production was zero in 1996 but was positive thereafter. In 1997, Qatar began exporting natural gas in the form of liquefied natural gas (LNG). Chapter 5 explains how liquefaction is beneficial for the transportation of natural gas. The density of LNG is about 600 times as high as that of natural gas in the form of vapor, although the density of the vapor is very

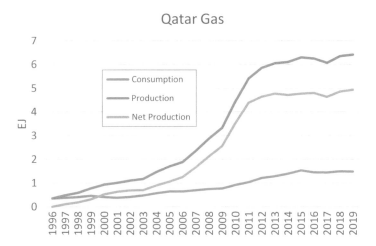

Fig. 3.12 Qatar natural gas production and consumption

sensitive to its pressure and temperature. Thus a tank or vessel of a given size can carry about 600 times as much natural gas as a liquid compared to a vapor.

Qatar exported 107.1 billion cubic meters of LNG in 2019, the largest LNG exports of any country. Natural gas that is liquefied and exported is not counted as consumption. So Qatar's very high per capita energy consumption is due to other factors besides LNG exports.

Gas is consumed in various large industrial facilities in Qatar. These facilities produce large amounts of ammonia, urea, ethylene, and methanol (see Table 3.4). Ammonia and urea are both important chemicals for use in agricultural fertilizers. It is energy intensive to produce them. Qatar has a ready supply of energy in the form of natural gas. Also, natural gas is used as a feedstock to provide the hydrogen necessary for the manufacture of ammonia, and it is also used as a feedstock for manufacturing urea.

Table 3.4 shows that Qatar also produces iron and steel. Qatar does not have reserves of iron ore, but the processing of iron ore into iron and steel is very energy intensive. Qatar's natural gas supplies energy for that processing.

Table 3.4 Qatar petrochemical and iron/steel production

Product	2016 Production (thousand metric tons)
Ethylene	829.7
Methanol	903.7
Ammonia	3424.1
Urea	5619.5
Basic iron and steel	6919.7

Natural gas can also be used as a feedstock to make clean-burning diesel, aviation fuel, lubricants, naphtha (a chemical intermediate used to make various products), and paraffin for detergents. This conversion process is known as gas to liquids (GTL). Qatar is home to the world's largest GTL plant, a joint venture with Shell called the Pearl GTL plant. It has the capacity to produce 260,000 barrels of GTL products and natural gas liquids per day.

Iceland

Table 3.1 shows that per capita energy consumption in Iceland is almost twice as high as that in Norway, even though both countries have similar climates. Again, industrial activity explains the difference. As of 2013 Iceland had three aluminum smelting plants with a total capacity of over 800 thousand metric tons per year. Aluminum smelting is the process of extracting aluminum metal from its oxide, alumina. The process uses large amounts of electrical energy. Iceland imports alumina for processing, and uses its large supplies of hydroelectricity and geothermal energy as energy sources. Iceland has several hydroelectric power stations (see Chap. 7 for a discussion of electricity). The largest is the Kárahnjúkar hydropower plant with a capacity of 690 MW. The large amounts of energy used for smelting result in large energy consumption in the country.

Trinidad and Tobago

Trinidad and Tobago are small Caribbean islands that produce large amounts of natural gas. The population of the islands is quite small (about 1.4 million people in 2019), so energy consumed in industrial processes on the island causes the per capita energy consumption to be high. Figure 3.13 shows natural gas production, consumption, and net production for the islands.

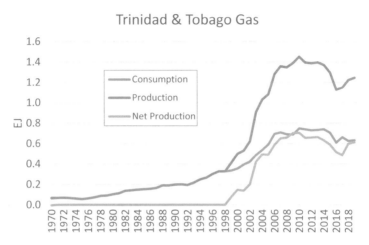

Fig. 3.13 Trinidad and Tobago natural gas production and consumption

The figure shows that all of the gas produced there was consumed until 1998. LNG exports began in 1999 from the Atlantic LNG plant. The plant currently has four trains with a total capacity of 15 million metric tons of LNG per year. As in Qatar, gas that is liquefied is not counted as being consumed in the country.

Gas consumption is strongly influenced by the large petrochemical facilities on the islands. As explained in Chap. 5, it is difficult to transport large amounts of natural gas without gas pipelines. Since Trinidad and Tobago is remote from large population centers, it is not economical to build gas pipelines to demand locations. Instead large amounts of natural gas are used to manufacture ammonia and methanol, as in Qatar. There are 11 ammonia plants with a total capacity of 5.2 million metric tons of ammonia per year and 7 methanol plants with a total capacity of 6.6 million metric tons per year. Gas used in these plants is included as consumption in the country. The country was the world's largest exporter of ammonia and the second largest exporter of methanol in 2013.[9] Both Qatar and Trinidad and Tobago have monetized their natural gas reserves not only by exporting LNG but also by manufacturing petrochemicals using the gas as a feedstock.

---

[9] https://en.wikipedia.org/wiki/Economy_of_Trinidad_and_Tobago

Singapore

Singapore is a major industrial center in Southeast Asia. About a quarter of its 2019 total industry revenue was generated by chemicals and refining.[10] The country is a global chemical hub. Although it has no reserves of oil or gas, it established itself as a major location for integrated refineries and chemical plants by reclaiming land to create the Jurong Island industrial park. Jurong Island hosts more than 100 refining and chemical companies, including some of the world's largest. Figure 3.14 shows some of the facilities on the island. Similar to Qatar and Trinidad and Tobago, feedstocks for the chemical plants are part of the explanation for the high per capita energy consumption.

Fig. 3.14 Shell's Pulau Bukom island site

But there is another factor which is even more important for understanding energy use in Singapore: marine and aviation bunkering, i.e., providing fuel for ships and airplanes, as shown in Table 3.5.

Table 3.5 Oil and product use in Singapore (2017)

PJ	Primary oil	Oil products	Total
Imports	2469	5017	7486
Exports	−54	−4296	−4349
International marine bunkers	0	−2051	−2051
International aviation bunkers	0	−332	−332
Total	2416	−1662	754

Positive numbers indicate imports; negative numbers indicate exports and use. Source: United Nations Statistics Division

The table shows that Singapore imported about 7.5 EJ of primary oil (crude oil, natural gas liquids, and other hydrocarbons) and oil products (such as feedstocks and naphtha). The country also imported a small amount of natural gas through a subsea pipeline from Indonesia. The oil and products provided inputs for the refineries (see Sect. 4.6 for a discussion of refining) and chemical plants. The country

---

[10] Department of Statistics, Singapore.

exported about 4.3 EJ, almost entirely as products (such as gasoline, aviation fuel, diesel, and residual fuel oil). Thus the net imports were about 3.1 EJ. About 2.1 EJ was used for marine bunkering and 0.3 EJ for aviation bunkering, leaving about 0.8 EJ for other consumption in the country, which includes feedstocks for chemical plants.

Sources

General: [1] BP Statistical Review of World Energy, June 2020

Figure

Number	Source
3.3	US Energy Information Agency
3.7	US Energy Information Agency
3.8	US Energy Information Agency
3.11	IEA (2018). The Future of Petrochemicals. All rights reserved

Tables

Number	Source
3.3	https://www.eia.gov/energyexplained/use-of-energy/transportation-in-depth.php
3.4	Qatar Planning and Statistics Authority
3.5	United Nations Statistics Division

Oil

4

4.1 Petroleum Sources

4.1.1 Subsurface Reservoirs

Oil is found in the subsurface of the earth, hundreds or thousands of feet below the surface. Oil reservoirs contain a mixture of many types of hydrocarbons and may include non-hydrocarbons such as carbon dioxide, nitrogen, and/or hydrogen sulfide (H_2S). Reservoirs with hydrogen sulfide are known as sour. Since H_2S is very toxic, extreme care is required for dealing with sour reservoirs. Pressure and temperature increase with depth below the surface, so reservoirs are at higher pressures and temperatures than the surface. Reservoirs occur both below land and also offshore in the subsurface below lakes and oceans.

Hydrocarbons in the subsurface are formed over geological time, starting with deep burial of organic matter. The high pressure and temperature result in the formation of hydrocarbons of various types. There are two types of petroleum reservoir: conventional and unconventional.

4.1.2 Conventional Reservoirs

Even though the subsurface of the earth may appear to be solid consisting of rocks and dirt, it actually contains water in pore spaces. Since the density of hydrocarbons is less than that of water, hydrocarbons tend to rise upwards, floating above the water. "Floating" is something of a misnomer, since the fluids in the subsurface are contained in rocks. Rocks may appear to be solid, but most rock types have pore space that contains fluids.

Under special geological circumstances, namely porous reservoir rock with the right geometry and an impermeable rock layer above it, the hydrocarbons can become trapped. There are two types of trap: anticline and fault. Figure 4.1 shows a cross section of an "anticline." The surface of the earth is above the region shown in

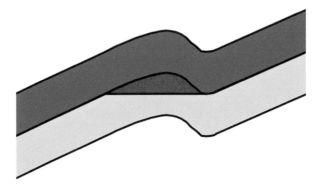

Fig. 4.1 Anticline trap

the illustration; the image is looking at the reservoir from the side. The yellow zone is porous rock and the green zone is an impermeable rock layer. Over long times in history, geological processes resulted in a fold in the rock layers. After the folding occurred, hydrocarbons (shown in red) migrated up through the porous rock and became trapped by the fold. More hydrocarbons may have migrated upwards but they leaked off to the right of the red zone shown in the illustration. This diagram shows a trap that is filled to "spill point." If the lower level of the red zone was higher, then the trap would only be partially filled.

Figure 4.2 shows another type of trap, known as a "fault trap." As before, the yellow zone is a permeable rock, while the green zone is an impermeable rock layer. These layers would have been deposited on top of each other in geological time. But subsequently the layers were tilted and slid past each other along a fault, shown as the heavy black line. Faults can also be observed after earthquakes on the surface of the earth. The red shows hydrocarbons trapped by the fault. Some faults, such as the one in the figure, are sealing so the hydrocarbons become trapped, whereas others are not.

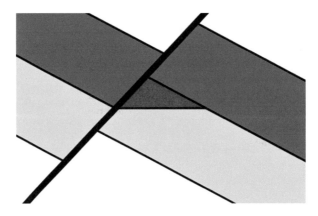

Fig. 4.2 Fault trap

Some conventional petroleum reservoirs have a distinct gas cap that floats above a zone with oil. Other reservoirs contain only gas. Most reservoirs have water below the hydrocarbons. A conventional reservoir requires several features: a trap such as an anticline or a fault that can contain the hydrocarbons, a seal that prevents them from escaping, a hydrocarbon source, and a migration path that allowed the hydrocarbons to flow from the source to the trap over geological time.

4.1.3 Unconventional Reservoirs

Unconventional reservoirs contain hydrocarbons that were formed within rock but never migrated. They have become increasingly important for hydrocarbon production in the twenty-first century. Shales are an example of an unconventional reservoir. Shales contain hydrocarbons in material with very low permeability. Permeability is a rock property that determines how easily fluids flow within them. Since shales have low permeability, hydrocarbons do not flow easily in them. However shales have been produced successfully by using a combination of two technologies: horizontal drilling over extended lengths within the shale and hydraulic fracturing to enable flow by splitting open the shale.

4.1.4 Tar Sands

There is another source of oil, called tar sands. Alberta Canada has enormous tar sand reserves. They are a mixture of bitumen (heavy oil) and sand. Tar sands are typically mined at the surface with earth-moving equipment: loaders and trucks. The tar sand is transported to a facility where it is heated with steam. This heating separates the bitumen from the sand. The sand is returned to the place where it was mined. The bitumen is a heavy oil that is very viscous so it does not flow well. It is processed in large plants called upgraders.[1] They change its chemical composition into shorter chain hydrocarbons that more closely resemble those in other crude oils.

Mining of tar sands requires earth moving on a huge scale (see Fig. 4.3). In addition, large amounts of energy are required to create the steam used to separate the bitumen from the sand. This energy is typically supplied by gas reserves that are located near the tar sands. There is no substantial alternative demand for that gas since the tar sands are located in remote locations with few people living nearby. Although it is technically possible to build a pipeline to bring the gas to a market, the remoteness of the location means that it would not be economical to do so.

---

[1] Some upgraders in Canada are located a significant distance south of the tar sands. There are dual pipelines to get the bitumen to the upgraders. One pipeline sends a diluent north to the tar sands, where it is blended with the bitumen to make diluted bitumen, also known as dilbit. The dilbit flows south to the upgrader in the other pipeline.

Fig. 4.3 Mining tar sand

4.2   Exploration

Since conventional and unconventional petroleum reservoirs are below the surface
of the earth, it can be challenging to find them. One possible method is to drill wells
at random locations. However, that would not make economic sense since it is costly
to drill wells. Ultimately, a well must be drilled to find a petroleum reservoir. But
the well locations should be chosen judiciously. Several methods are used to choose
drilling locations that have a reasonable probability of a reservoir below. Seeps of
oil or gas at the surface are an indication that there may be hydrocarbons below the
surface. However, since the migration path to the surface may not be vertical, the
reservoir may not be directly below the location of the seep.

Another exploration method is to look for geological structures on the surface
that are similar to the anticline illustrated in Fig. 4.1. One example would be a hill
at the surface. Alternatively, there can be something similar: sometimes a hill that
existed on the surface at a prior geological time is eroded away as shown in Fig. 4.4.
The brown horizontal line in the figure indicates the earth's surface. At the surface
the rock type shown in green is surrounded by another rock type shown in white.
Based on this, one can infer that an anticline may exist below the surface.

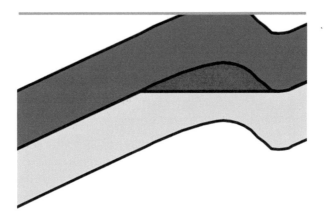

Fig. 4.4 Surface expression of an anticline

There are other sensing techniques that can be used to identify possible petro-leum reservoirs. Airplanes can be used with special equipment to measure gravity or magnetic anomalies below the surface. Another very common exploration tech-nique is "seismic." This may bring earthquakes to mind, but exploration seismic is different. Seismic involves the use of a sound source, dynamite or a thumper truck, and geophones arrayed on the surface. A geophone is a microphone that detects sound from the source that is reflected back from the subsurface to the geophones. The time delay between the sound generation and the receipt of the reflected sound at the geophones can be used to determine the distance traveled from the source to a reflecting layer below the surface and back to the geophone. Since each change in rock layer type below the surface generates a reflection of the sound, it is computa-tionally intensive to convert sound travel times to an image of the subsurface. Processing of raw seismic data requires substantial computing power, software, and expertise. Seismic can identify both anticlines and faults. Seismic is done both onshore and offshore. Offshore seismic uses boats with floating geophones.

Exploration drilling is risky even after doing studies such as seismic. In 1983, Sohio drilled the Mukluk exploration well in offshore Alaska to a depth of 9860 ft at a cost of $430 million.[2] Exploration activity prior to drilling indicated that a res-ervoir could contain at least 1.5 billion barrels of oil. The drilling was near the giant Prudhoe Bay Oil Field which had been producing oil since 1977. This all inspired enough confidence that Sohio and partners was willing to invest hundreds of mil-lions of dollars. It was a huge disappointment when tests of the well after drilling produced only salt water and a small amount of gas, but no oil. The well was plugged with cement and abandoned.

---

[2] Sohio had a 31.4% interest in the well. Other companies owned the remainder.

4.3   Drilling

Each of the exploration methods described above, except seeps, can detect traps, but there is no guarantee that a trap will contain hydrocarbons. The trap may only contain water if there was no hydrocarbon source, or if migration from the source occurred before formation of the trap. The only sure way to find a hydrocarbon reservoir is to drill a well. Figure 4.5 shows the cross-sectional schematic for a vertical well. The figure shows that the well is initially drilled with a large-diameter bit, large enough that a 30-in. steel pipe can be lowered into the top 20 m of the well (about 65 ft). This large-diameter pipe is known as a conductor. The conductor is cemented in place by pumping cement through the middle of the conductor and up its outside to make contact with the surrounding hole.

Fig. 4.5 Vertical well schematic

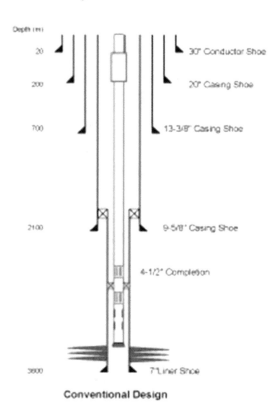

Conventional Design

Next a smaller bit is used to deepen the well, as in the case of the figure, to 200 m. A second steel pipe, known as casing, with a 20-in. diameter, is then lowered into the well and also cemented in place. Then a smaller bit is used to deepen the well yet further, as in the case of the figure, to 700 m. A third steel pipe, a 13 3/8-in. casing string in the figure, is then lowered into the well and cemented in place. Next a smaller drill bit is used to deepen the well to 2100 m in the case of the figure. A fourth steel pipe, a 9 5/8-in. casing string, is lowered into the well and cemented in place. Finally, in the case of the figure a yet smaller bit is used to drill down to 3800 m.

Measurements with sophisticated logging tools can be made while a well is being drilled. These well logs provide data about rock properties near the well and the presence or absence of hydrocarbons. If logging indicates the presence of hydrocarbons, then a flow test can be conducted to determine (1) the ability of the well to produce hydrocarbons, (2) the type of hydrocarbon (oil, gas, or a combination), and (3) the flow rate that may be expected from the well.

A flow test is conducted when the well reaches a hydrocarbon zone. If the well is productive, then a final 7-in. casing string, known as a liner, may be lowered into the well, and hung from the 9 5/8-in. casing string by an annular hanger, shown with a black cross in the figure.[3] The final casing string is also cemented in place. If the well is not productive, then cement is pumped down to prevent any flow occurring from the well.

When all the casing strings are in place, the liner is perforated at the level of the reservoir using shaped charges that are lowered into the well. This blasts holes in the liner and creates fractures in the rock outside the well (shown as red streaks at the bottom of the well in the figure) to allow hydrocarbons to flow into the well. The hydrocarbons are produced through a tubing string, shown as the innermost pipe with a diameter of 4 ½ in. in the figure. The red cross-hatched item between the tubing string and the liner is known as a packer. The packer prevents hydrocarbons from flowing up the annular space between the tubing string and the innermost casing string and/or liner.

This may seem like an unnecessarily complicated design to reach a reservoir below the surface. Not all wells have the same design as shown in the figure, but all wells will be somewhat similar. Note that the hydrocarbons are produced from the innermost tubing string, and that there are three layers of steel pipe between the hydrocarbons and the surrounding earth from 200 to 700 m below the surface. There is an additional steel layer from 20 to 200 m, and yet another layer from the surface to a depth of 20 m. These many layers of steel provide substantial protection for any aquifers that lie outside the well below the surface of the earth.

The discussion above has gone into some detail about the subsurface details of a well. Many people believe that all oil wells look something like Fig. 4.6. Indeed there are many oil wells like this. The device in the figure is known as a pump jack. It is one example of artificial lift, which is used late in the life of a well to increase the production rate. The surface facilities such as this pump jack do not reveal much about the design of the well below the surface.

---

[3] A hanger is not always used. Alternatively, an additional casing string may be used from the surface all the way down to the reservoir.

Fig. 4.6 Pumping oil well

Artificial lift is not used early in the life of a well, since wells can flow naturally to the surface driven by the high reservoir pressure. A typical wellhead without artificial lift is shown in Fig. 4.7. The wellhead has several valves to control flow from the tubing string, as well as to enable access to the annulus, i.e., the region

Fig. 4.7 Wellhead

between the tubing and the casing. This access is useful for various maintenance operations. Quaintly, the industry refers to such a wellhead as a Christmas tree. The valves are not the most ornate ornaments you have ever seen!

So far the discussion in this section has been for a vertical well. As mentioned above there are also horizontal wells, such as shown in Fig. 4.8. Horizontal wells are used for production from shales and also in some conventional reservoirs when an extended contact between the well and the reservoir is desired, often due to low rock permeability.

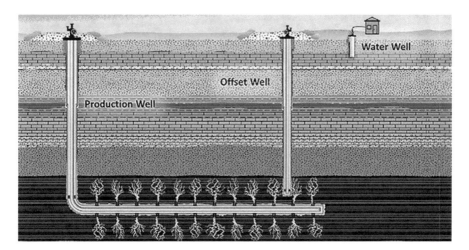

Fig. 4.8 Horizontal well

The figure shows two wells contacting the reservoir, one vertical and the other horizontal. Often one or more vertical wells are drilled into a reservoir before a horizontal well is drilled. The figure also shows shallow aquifers in blue. Aquifers are protected by using multiple casing strings such as shown in Fig. 4.5 in both horizontal and vertical wells.

The vertical offset well shown in the figure has only a limited contact with the reservoir, whereas the horizontal well contacts the reservoir over the entire horizontal section. The horizontal section of some wells is as long as 3 miles! The figure also shows jagged structures emanating from the horizontal section. This represents the result of hydraulic fracturing, which is used to enable flow in shale or other rock that has low permeability. Fracturing is achieved by pumping large quantities of water at high pressure into the well together with sand and a small amount of chemicals. The high-pressure water splits open the rock. The sand flows into the fractures in the rock, keeping them open once the injection stops. The chemicals serve various purposes such as ensuring that the sand is carried with the water. Horizontal wells have many fractures. Figure 4.8 shows a well with 12 fracture stages. There are clever mechanical devices that enable sequential fracturing in different parts of the horizontal section.

The large volume of water injected at high pressure requires an impressive amount of equipment, such as shown in Fig. 4.9. The trucks in the figure carry

Fig. 4.9 Hydraulic fracturing equipment

pumps. The array of hoses is used to carry the hydraulic fracturing fluid. A frac job is usually finished within a few hours. Many oil field operators have small pads from which they drill multiple horizontal wells. It is common for several wells to be drilled from a pad, followed by fracking of those wells. Fracking is one type of well "completion." The fracking equipment leaves the well site after the wells are completed, in order to perform similar operations at other well sites.

Some oil companies have an inventory of drilled but uncompleted wells (DUCs). The cost of a completion for a horizontal well is often almost as much as the cost of drilling the well. Completion may be delayed if the value of hydrocarbons falls while drilling to such an extent that it would not be profitable to do the completion, or if there is a lack of pipeline capacity to remove produced hydrocarbons from the wells.

After a frac job has been completed, the water that was injected flows back to the surface. The flowback water is usually too salty to be used for irrigation or other purposes, so it is reinjected into the subsurface deep below aquifers. There are indications that this reinjection, together with the injection of wastewater that is produced during well operations, triggers small earthquakes. There was a notable increase in seismicity (earthquakes) in Oklahoma starting in 2009. Figure 4.10 shows that there were more earthquakes with magnitude 3 or greater in Oklahoma than in California from 2014 to 2017. Such earthquakes are surely noticeable, but small compared to earthquakes that cause massive damage. For example, the magnitude of the 1989 Loma Prieta earthquake that caused extensive damage in northern California was 6.9. The Richter scale that is used to measure the magnitude of earthquakes is logarithmic, so magnitude 6.9 is 1000 times stronger than magnitude 3.9.

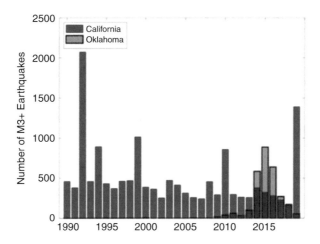

Fig. 4.10 Seismicity in California and Oklahoma

4.4   Production

4.4.1   Overview

If a conventional reservoir is discovered then one or more appraisal wells are drilled nearby to assess the size of the discovery[4] and likely production rates of oil and gas.[5] Economic forecasts are made based on estimates of the number of wells to be drilled, the cost to drill the wells and to construct surface facilities to produce the hydrocarbons, the anticipated production rates, the operating costs, and the prices anticipated to be received for oil and gas. Since oil fields typically produce for decades, these forecasts are subject to considerable uncertainty. If the economics look attractive, then the reservoir is developed. This means designing and building surface facilities, designing, drilling and completing wells, and often building infrastructure such as pipelines to bring the hydrocarbons from the field to a market.

In the case of an unconventional reservoir discovery, appraisal is not necessary. But considerable effort is devoted to optimizing the design of each well including the length of the horizontal section, the spacing between wells, and the specifics of hydraulic fracturing such as the number of fracture stages.

---

[4] The reservoir size can also be estimated from a single well using a technique known as pressure transient analysis. The well is produced for a period of time while the flow rate and pressure are measured. The well is then shut in and the pressure is monitored. The decline in pressure during the production phase and the subsequent buildup of pressure when the well is shut in can be used to estimate the size of the reservoir.

[5] Hydrocarbon wells produce oil, gas, and water.

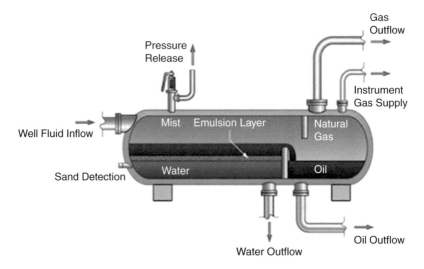

Fig. 4.11 Three-phase separator

Oil wells do not produce only oil. They also produce gas and water. The fluids that are produced in a well[6] are sent to a separator, which has three output streams: oil, gas, and water as shown in Fig. 4.11.

Fluids from a well or group of wells enter at the left of the separator. Gas, being the least dense fluid, exits at the top of the separator. Oil and water exit at the bottom of the separator. Note that oil floats on a water layer at the bottom left of the separator. Oil floats on water since it is less dense than water. There is an emulsion layer consisting of oil and water together between the oil and water layers. The oil pours over a weir inside the separator, resulting in an oil phase at the bottom right of the separator that is mostly free of water and gas.

4.4.2   Natural Depletion

All reservoirs are produced by drilling wells into them and bringing hydrocarbons to the surface. Flow into a well occurs because the pressure at the bottom of the well is lower than the pressure in the reservoir. Fluids move from a region of high pressure to a region of low pressure.[7] As fluids leave the reservoir over time the reservoir pressure drops, unless there is water influx or injection of other fluids. Influx and injection are described below.

Lower pressures in the reservoir result in lower flow rates to a well, since there is less driving force. Thus the production rate from wells declines over time, unless

---

[6] Typically fluids from several wells are collected in a series of surface pipes known as a gathering system. The gathering system delivers the fluids to one or more separators.

[7] The flow rate depends both on the pressure gradient and on the permeability. For the same pressure gradient, the flow rate is higher if the permeability is higher.

steps are taken to reverse this natural decline. Wells that produce based solely on reservoir pressure are said to be undergoing "primary depletion."

Many hydrocarbon reservoirs are in contact with a water source, which is continuously replenished by naturally occurring water if fluids are removed from the reservoir. Depending on the strength of the water replenishment, the decline in reservoir pressure is moderated, and may even stay roughly constant. This pressure support can enhance production rates in a well, but it also has a downside. Water influx to the reservoir can also flow into the well. Thus the well will produce a mixture of oil and water. Also, since the density of water is higher than the density of oil (except for some very heavy crude oils) the pressure at the bottom of the well will increase as the column of liquid in the tubing string contains more water. Higher pressure at the bottom of the well means there is less of a pressure drop from the reservoir to the well. Typically the water production rate in oil wells increases over time relative to the oil production rate. The ratio of water production to the total liquid production is known as the "water cut."

The rate of pressure decline in a reservoir can be moderated by another factor. An oil reservoir may have a region of gas above it, known as a gas cap. Even if there is not a gas cap initially, a gas cap typically forms when the pressure in an oil reservoir becomes low enough that gas comes out of solution. You have probably seen gas come out of solution when you open a bottle of soda. Before you open the cap there is mostly liquid in the bottle with a small apparently empty space above the liquid. When the cap is opened, the pressure in the bottle is reduced and gas that was dissolved in the liquid comes out of solution in the form of bubbles. The same thing happens in an oil reservoir when the pressure drops below a certain threshold, known as the bubble point. The bubble point pressure depends on the composition of the crude oil and on the temperature in the reservoir. The gas bubbles that form in a reservoir float to the top and also flow into the well.

Unlike liquids, gas is quite compressible. When its pressure increases its volume decreases, or conversely gas expands if its pressure is reduced. The expansion of a gas cap as reservoir pressure falls can slow the decline in pressure as fluids are removed from the reservoir.

Natural depletion is also referred to as "primary recovery." Only about 10% of the oil that is originally in place in conventional reservoirs is recovered with natural depletion. Since so much oil is left behind, other methods (described below) are used to increase the amount of oil that is recovered over time.

4.4.3 Multiphase Flow

As mentioned above, reservoirs usually produce three types of fluid: oil, gas, and water. All of these fluids flow through narrow pore spaces in the formation. The relative flow rates of each of these fluids depend on surface tension and viscosity. Since gas has a much lower viscosity than oil or water, it tends to flow more easily than the liquids. Over time, the gas-oil ratio, i.e., the ratio of the gas flow rate to the oil flow rate, tends to increase. Also, when there is pressure support from an aquifer, the water-oil ratio tends to increase over time.

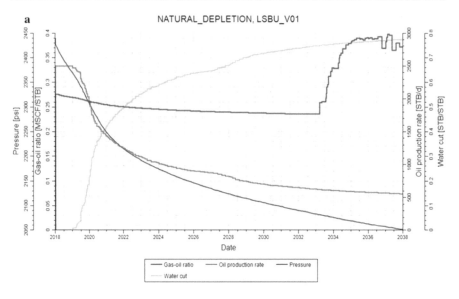

Fig. 4.12 Oil, gas and water production

Multiphase flow for a well is illustrated in Fig. 4.12. There is a lot going on in this figure! The horizontal axis is time, going from 2018 to 2038. It is common for wells to produce for long periods like this. The green curve shows how the oil production rate varies over time. The scale for the rate is shown on the right-hand side of the figure, and is measured in "STB/d," which means stock tank barrels per day. A stock tank barrel is a quantity of oil at standard conditions of pressure and temperature.[8] The oil rate starts at about 2500 STB/d, and remains constant for about a year. After that the green curve drops, indicating that the oil production rate falls. The drop begins to occur at the same time as the blue curve, which represents the water cut, and increases above zero. The water production rate increases over time as more and more water flows into and up the well. The oil production rate falls as the water cut (and water production rate) increases.

The brown curve shows how the pressure falls over time as fluids are removed from the reservoir. The relatively rapid drop in pressure indicates that there is little if any pressure support from an aquifer in this case. The red curve shows the ratio of gas production to oil production measured in MSCF[9] gas per STB of oil. This is

---

[8] The density of oil is different at higher pressures and temperatures that exist in the reservoir than in a sock tank at the surface. The density is also affected by gas that is dissolved in the oil at reservoir conditions.

[9] 1 MSCF is 1000 standard cubic feet, i.e., the amount of gas contained in 1000 cubic feet at standard conditions of pressure and temperature, 60 °F, and atmospheric pressure at sea level.

known as the gas-oil ratio (GOR). For the first 15 years, the GOR falls over time. This is because gas is coming out of solution in the reservoir as the pressure falls. So less gas is contained in the oil that is produced. However, in 2033 the GOR will begin to increase significantly. After this time, large quantities of gas will be flowing into and up the well. One can infer that a gas cap has been forming in the reservoir before this time. After 2033, free gas wil be flowing from the gas cap to the well. Since the viscosity of gas is much lower than that of liquids, the gas flow rate can be quite high.[10]

4.4.4 Artificial Lift

When the pressure at the bottom of a well is lower, there is more driving force for fluids to flow from the reservoir to the well. Thus production rates can be increased by lowering the pressure at the bottom of a well. The pressure at the bottom of a well can be lowered by installing a pump. The pump shown in Fig. 4.6 is a sucker rod pump that is commonly used in oil wells. The motor at the left of the figure causes the horsehead at the top of the pump to rock up and down. Fluid is brought into and up the well on the upstroke. This is like a hand pump that is used for water wells.

An alternative to a sucker rod pump is an electric submersible pump. These are centrifugal pumps that are lowered to the bottom of a well and powered by an electrical cable from the surface.

There is yet another method of artificial lift known as gas lift. When a well is producing there is a column of liquid in a well. If you have ever done scuba diving, or simply dived to the bottom on a swimming pool, you will have experienced the pressure increase that occurs with depth as you go deeper. Similarly, the column of liquid in a well results in high pressures at the bottom of a well. To counteract this, gas can be injected into the annular region between the tubing string and the casing, and then passed through a valve into the tubing string. This creates bubbles in the tubing string (see Fig. 4.13), reducing the density of fluids in the tubing. It is the density of the fluid that causes pressure increase with depth. If the pressure at the top of the well is the same, then the lower density caused by gas in the tubing lowers the pressure at the bottom of the well. In summary, sucker rod pumps, electrical submersible pumps, and gas lift are different methods of artificial lift.

---

[10] There is a concept known as "relative permeability" that describes the relative flow rates of different phases (oil, gas, and water) in a reservoir.

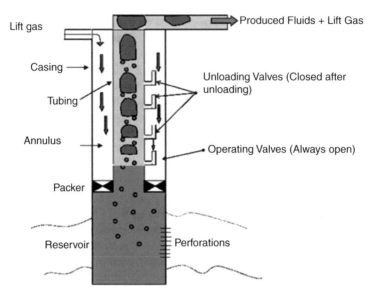

Fig. 4.13 Gas lift

4.4.5   Secondary Recovery

As mentioned above, the pressure in an oil reservoir typically falls as fluids are removed. There is an exception if the oil reservoir is connected to a water aquifer than provides pressure support. In a similar manner to aquifer pressure support, it is possible to add fluids to a reservoir that is being produced in order to limit the rate at which reservoir pressure falls. This is known as "secondary recovery." Either gas or water can be injected into the reservoir to support reservoir pressure. Injection wells are drilled for this purpose some distance from the producing wells. If gas is used, it is injected near the top of the reservoir which will often have a gas cap anyhow due to the lower density of gas relative to liquids. If water is used, it is injected near the bottom of the reservoir, since the higher density of water tends to make it sink to the bottom. Water injection is known as waterflood. The injection wells are located some distance from the producing wells in order to reduce the tendency of the injected fluids to flow towards the producing wells. The purpose of secondary recovery is to produce more oil, not to simply circulate injection fluids from an injection well to a production well. Unfortunately, reservoirs tend to be heterogeneous with a few high permeability layers. Since the pressure at the bottom of an injection well is high while that at the bottom of a producing well is low, the injection fluids often flow through these high-permeability layers, bypassing the oil in the remainder of the reservoir and causing circulation of the injection fluids. This circulation is referred to as breakthrough. There are technical methods for reducing this breakthrough tendency.

Application of primary and secondary techniques achieves about 20–40% recovery of the oil originally in place in conventional reservoirs.

4.4.6 Thermal Oil Recovery

Some crude oil is heavy, dense, and viscous like tar. Such oil does not flow readily in a reservoir. Viscosity can be reduced by heating the oil. Since reservoirs are deep below the surface, it is not possible to place heaters in the subsurface.[11] One method for heating the oil is to inject steam into the reservoir. Heat from the steam tends to be lost to the surrounding formation in injection wells, so this method is restricted to relatively shallow wells. There are two approaches to steam injection: steam flood, which is similar to a waterflood, or cyclic steam injection. With cyclic injection steam is injected for a period of time to heat the oil in the vicinity of the injection. Then flow in the well is reversed, so that it becomes a producing well. Steam injection is relatively common in parts of California. Steam injection requires an energy source to convert liquid water to steam. Natural gas is commonly used as the fuel for steam generation. Efficiency of steam generation can be increased substantially by using cogeneration. A cogeneration plant uses a fuel such as natural gas to generate electricity (as discussed in Sect. 7.2.2). Such plants produce steam as a by-product, which can be used for injection into the oil reservoir.

4.4.7 Tertiary Recovery

As shown in Fig. 4.14, oil in a reservoir is contained in small-pore spaces within a rock matrix. Surface tension tends to trap the oil in these pore spaces, even when an oil field is produced.

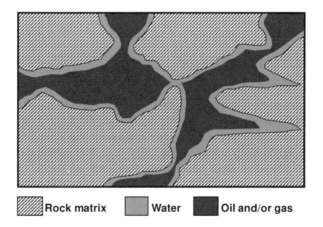

Rock matrix   Water   Oil and/or gas

Fig. 4.14 Oil in pore space. Medo Hamdani

---

[11] Some companies did develop technologies that could be used to heat the subsurface using electrical heating. The power requirements were enormous, and thus this technology was not economic.

There are so-called tertiary recovery techniques that overcome such surface tension effects, and allow additional oil to be recovered from conventional reservoirs. These techniques are also known as enhanced oil recovery. They all involve injection of a fluid into the reservoir that affects the physical properties of the oil. One such injectant is carbon dioxide; another is a gas mixture including methane, ethane, and propane that becomes miscible with the oil at reservoir pressure and temperature. This miscibility allows for additional oil recovery. Sometimes there is alternating injection of a tertiary fluid and water to improve oil recovery. It is also possible to use polymers or detergent-like surfactants as injectants for tertiary recovery. There is an additional benefit from the use of carbon dioxide, namely the carbon dioxide is captured in the subsurface rather than being released into the atmosphere.[12]

About 30 to 60 percent of the oil originally in place in conventional reservoirs can be recovered after application of tertiary techniques. This means that about 40 to 70 percent of the original oil remains in an oil reservoir after it is produced even after tertiary recovery. If the price of oil becomes high enough, then techniques will likely be used to recover some of this oil. One such technique is to drill new infill wells to recover additional oil.

4.5   Transportation

Crude oil that is produced from wells requires processing into finished products such as gasoline, diesel, and jet fuel. Such processing occurs in refineries that are discussed below. So the oil must be transported from oil fields to refineries. Four methods are used for transportation: pipelines, shipping in oceangoing tankers and barges, trucks, and railcars.

4.5.1   Pipelines

There are both land and marine crude oil pipelines. Almost all land pipelines are buried at shallow depths in the ground. Pump stations are required to provide motive force for oil flow in the pipelines. They are located on the surface. Marine pipelines may lay on the seafloor or be buried in a trench dug into the seabed. There are about 72,000 miles of crude oil pipelines in the USA. A map of these pipelines is shown in Fig. 4.15. Notice the large number of pipelines in Texas, Oklahoma, and Louisiana. These are states with substantial oil production. Oil pipeline diameters

---

[12]This storage benefit is for carbon dioxide from combustion sources like power plants. Some carbon dioxide is itself produced from underground reservoirs that contain carbon dioxide, such as Sheep Mountain in Colorado.

Fig. 4.15 US (Lower 48) crude oil pipelines

range from 4 in. to 48 in.[13] About 70% of crude oil and petroleum products in the USA are transported by pipelines.[14]

> The Trans-Alaska Pipeline System (TAPS) is an 800 mile long 48 in. diameter crude oil pipeline from inside the Arctic Circle in the north of Alaska to a marine terminal in Valdez, located on the south coast of Alaska. Very unusually about half of this pipeline is aboveground to protect the cold permafrost from the hot oil. At its peak, about two million barrels of oil per day flowed through the system. TAPS is notorious for the oil spill that occurred from the Exxon Valdez oil tanker in 1989.

4.5.2 Water Transport

Large quantities of crude oil are shipped worldwide in tanker ships such as shown in Fig. 4.16.[15] Oil shipping accounts for nearly a third of all global maritime trade.[16] Table 4.1 shows the different size classes for oil tankers. The size classes Suezmax

---

[13] Oil may be transported from individual wellheads to central processing sites in gathering lines with smaller diameters.

[14] https://www.forbes.com/sites/jamesconca/2014/04/26/pick-your-poison-for-crude-pipeline-rail-truck-or-boat/#4a99884117ac

[15] This particular tanker was hijacked by Somali pirates in 2008. The ship was released after payment of a $3 million ransom.

[16] https://www.planete-energies.com/en/medias/close/transporting-oil-sea

Fig. 4.16 VLCC

Table 4.1 Classes of marine oil tanker

Tanker class	Maximum deadweight tonnage	Capacity barrels
Ultra-large crude carrier (ULCC)[a]	500,000	4,000,000
Very large crude carrier (VLCC)	300,000	2,000,000
Suezmax	200,000	1,000,000
Aframax	120,000	750,000
Panamax	80,000	500,000

[a]Most ULCCs have been broken up or converted to other purposes

and Panamax are defined by the maximum dimensions of vessels that can traverse the Suez Canal and the Panama Canal, respectively.

Barges are used for water transport of smaller volumes. Barges are not self-propelled, but require tugboats instead. About 23% of crude oil and petroleum products are transported by water in the USA.[17]

---

[17] https://www.forbes.com/sites/jamesconca/2014/04/26/pick-your-poison-for-crude-pipeline-rail-truck-or-boat/#4a99884117ac

4.5.3 Trucks

It is costly to build pipelines. For small oil production rates it is more cost effective to store oil in tanks on the surface near one or more producing wells, and to remove it from the tanks by trucks on a regular schedule. Only about 4% of US crude oil is transported in this way.[18]

4.5.4 Rail

Railways are also used to transport crude oil. During 2018 about 201 million barrels of crude oil were moved by rail in the USA.[19] Of this, 114 million barrels were movements within the USA and 87 million barrels were imports from Canada. Intra-US movements by rail in 2018 were about 3% of US crude oil production. Rail is a convenient way to transport substantial volumes of crude oil from regions which do not have pipeline infrastructure. For example, crude oil production from the Bakken Formation in North Dakota increased enormously after 2006. Production was about 2000 barrels per day in 2006, and rose above 200,000 barrels per day by 2010, and to almost 1.4 million barrels per day by mid-2019. Rail was used to transport substantial quantities of crude oil from the Bakken until oil pipelines were constructed. The controversial Dakota Access Pipeline began flowing oil from the Bakken in 2017.[20]

Unfortunately, there have been accidents involving crude oil trains. Forty-seven people died in Lac-Mégantic, Canada, in July 2013 when a train derailed. There have been other accidents in the USA involving oil transport by rail.

4.6 Refining

4.6.1 Products and Inputs

Crude oil produced from wells is a mixture of various hydrocarbons. This mixture is not suitable for use in its raw form, but must be processed into products that are suitable for various purposes. Table 4.2 shows the range of products that were produced in US refineries and blenders[21] in 2019. About half of the output was finished gasoline; about a quarter was distillate fuel oil. Distillate fuel oil includes diesel fuel

---

[18] https://www.forbes.com/sites/jamesconca/2014/04/26/pick-your-poison-for-crude-pipeline-rail-truck-or-boat/#4a99884117ac

[19] https://www.eia.gov/dnav/pet/pet_move_railNA_dc_ZAMN-ZAMN_mbbl_a.htm

[20] Protests about potential spills from the pipeline and interference with Native American sites caused lengthy delays in start-up of the pipeline. In July 2020 a court ruled that the pipeline should be shut down and emptied. The ruling was overturned in August 2020, and legal challenges will likely continue.

[21] The distinction between refineries and blenders is discussed below.

Table 4.2   US refinery and blender net production, 2019

	Average daily production, thousand barrels per day	%
Finished motor gasoline	10,090	49.4%
Distillate fuel oil	5,136	25.1%
Kerosene-type jet fuel	1,796	8.8%
Petroleum coke	850	4.2%
Still gas	669	3.3%
Hydrocarbon gas liquids	606	3.0%
Residual fuel oil	363	1.8%
Asphalt and road oil	322	1.6%
Petrochemical feedstocks	299	1.5%
Lubricants	167	0.8%
Other	146	0.7%
Total	20,444	100.0%

that is used in vehicles and fuel oil that is used in industrial plants and commercial burners. About 9 percent of refinery output was jet fuel.

Output includes lesser amounts of several other products. Petroleum coke is a solid product that is used for fuel in electric power plants and cement kilns. It is a clean burning fuel that leaves virtually no ash when burned. Some petroleum coke is further processed to produce anodes that are used in the aluminum and steel industries. Still gas is primarily methane and ethane that is used as fuel in refineries or as a feedstock for petrochemicals. Hydrocarbon gas liquids are mixtures of alkanes, in particular ethane, propane, butane and pentane, and alkenes (or olefins). Those same alkanes are also produced in gas plants, where they are called natural gas liquids (see Chap. 5). They are used as feedstock for petrochemicals and fuels, as additives for motor gasoline, and as a diluent for transportation of heavy crude oil.

Residual fuel oil, also known as No. 5 and No. 6 fuel oil, is a heavy oil that is used in maritime vessels,[22] for electric power generation, heating, and other industrial purposes. Asphalt, also known as bitumen, is a tar-like substance that is mixed with gravel for road construction, and is also used for sealing roofs. Refineries also produce various lubricants such as engine oil.

Crude oil is by far the main input to refineries as shown in Table 4.3. However there are other inputs including renewable fuels (primarily ethanol produced from corn), hydrocarbon gas liquids from gas plants, motor gasoline blending components from other refineries, unfinished oils, and hydrogen. Unfinished oils are hydrocarbon mixtures, including naphtha, kerosene, light and heavy gas oils, and residuum, which is a heavy stream produced in refineries.

You may have noticed that the total output shown in Table 4.2 is larger than the total input in Table 4.3. This is typical and is referred to as refinery gain. The units

---

[22] The marine sector consumes about half of all global fuel oil. Effective January 1, 2020, the International Maritime Organization ruled that sulfur emissions from fuel oil must be reduced by 80%. This has implications for refineries that produce fuel oil. https://www.woodmac.com/nslp/imo-2020-guide/

Table 4.3 Inputs to US refineries and blenders, 2019

Input	Average daily input, thousand barrels per day	%
Crude oil	16,562	85.5%
Renewable fuels (including fuel ethanol)	981	5.1%
Gasoline blending components	684	3.5%
Hydrocarbon gas liquids	571	2.9%
Unfinished oils	355	1.8%
Hydrogen	213	1.1%
Total	19,366	100.0%

in each of the tables are volumes. According to the laws of science, mass is con-served but not volume. The average density of the products is less than that of the inputs. Thus more barrels of product are produced than the barrels of input.

4.6.2   Refining and Blending

As mentioned above, there are two broad types of oil processing: refining and blend-ing. Table 4.4 shows the net flows for each that occurred in the USA during 2019. A "net" flow is the difference between an input and an output (or product). A negative

Table 4.4 Net flows for US refineries and blenders, 2019 (thousand barrels per day)

	Refiners	Blenders	Total
Net inputs			
Crude oil	−16,562	0	−16,562
Natural gas liquids (butanes and natural gasoline)	−479	−92	−571
Renewable fuels (fuel ethanol and biodiesel)	−71	−910	−981
Hydrogen	−213	0	−213
Unfinished oils	−351	0	−351
Gasoline blending components	6,826	−7,512	−686
Subtotal	−10,850	−8514	−19,364
Net outputs			
Finished motor gasoline	1,602	8,446	10,048
Distillate fuel oil	5,078	31	5,109
Kerosene-type jet fuel	1,803		1,803
Petroleum coke	850		850
Still gas	669		669
Natural gas liquids (ethane, propane, butanes)	329		329
Refinery olefins	277		277
Residual fuel oil	363		363
Asphalt and road oil	322		322
Petrochemical feedstocks	299		299
Lubricants	167		167
Other	146		146
Subtotal	11,904	8477	20,382
Total	1,054	−37	1,018

Positive = net output, negative = net input
There are minor discrepancies compared to Table 4.2 and Table 4.3, due to use of different EIA tables

number in the table indicates a net input, while a positive number indicates a net output. For example there was a net input of 16,562 thousand barrels per day of crude oil to refineries during 2019 and no net flow of crude oil at the blenders.

The blenders had two significant net inputs: mostly motor gasoline blending components and secondarily some renewable fuels. These provided about 8.5 million barrels per day in total to the blenders. Virtually all of the output of the blenders was finished motor gasoline, with a very small amount of distillate fuel oil, hence the name of their operation. The blenders combine gasoline components to produce finished gasoline with appropriate specifications, such as octane rating, vapor pressure, and limits on emissions of carbon monoxide and volatile organic compounds after combustion. The bottom-line total net flow for the blenders was almost zero, indicating that the volume of products was almost exactly the same as the volume of inputs.

The situation with the refiners was quite different. They consumed mostly crude oil together with other inputs such as natural gas liquids, unfinished oils, hydrogen, and some renewable fuels to produce motor gasoline blending components, distillate fuel oil, jet fuel, some finished gasoline, and several other products as shown in the table. In contrast to blenders, refiners had a small net refinery gain of about one million barrels per day.

The US blenders consumed about 686,000 barrels per day more motor gasoline blending components than was produced by US refineries. This excess was imported from non-US sources. Similarly, the refineries imported unfinished oils, primarily heavy gas oils, from non-US sources.

Notice that the refineries produced virtually all of the distillate fuel oil, jet fuel, and petroleum coke, while the blenders produced hardly any of these products. Both the refineries and the blenders produced finished motor gasoline, but the bulk of this was produced by the blenders.

4.6.3   Refining Technology

As mentioned above, crude oil is a mixture of many different compounds, most of which are hydrocarbons. This mixture is not useful in its raw form. The compounds must be separated and many of them must be processed into useful products like gasoline. The main technique that is used to separate the compounds is distillation. All chemical compounds can exist in various physical states: gas (i.e., vapor), liquid, or solid. For example, ice is solid water, drinking water is liquid, and steam is gaseous water. The transition between liquid and vapor is known as boiling. You have seen this when water boils in a kettle or in a saucepan.

The temperature at which a pure compound boils depends on the compound itself and on the pressure. The boiling temperature is known as the boiling point. If you heat a mixture of compounds such as crude oil in a container, then the compound with the lowest boiling point boils first and becomes vapor. Compounds with low molecular weights tend to have low boiling points (see Table 4.11 in the appendix to this chapter). The vapor rises up out of the liquid mixture and can be collected as a relatively pure stream.[23] When the compound with the lowest boiling point has

---

[23] There are some exceptions to being able to capture a pure stream with only one compound. These exceptions are explored in physical chemistry classes.

completely evaporated from the mixture then the temperature rises until another compound boils and rises above the mixture. Thus pure streams of various compounds can be recovered from the mixture over time. This process of heating a mixture in a container and collecting the streams over time is known as batch distillation. It is used in labs, and for making liquor such as brandy or scotch.

Batch distillation is not suitable for separating large quantities of liquid mixtures, since containers must be filled, boiled over time, emptied of the heavy residue that remains, and then refilled as the process repeats itself. There is a continuous form of distillation that overcomes the non-steady-state nature of batch distillation. Figure 4.17 shows a schematic of a continuous fractionating distillation column such as those used in oil refineries. The liquid mixture of hydrocarbons enters as feed on the left side of the figure in the middle of the column. The liquid falls on to a feed tray. The pale blue upward arrows show the path of vapor from the feed tray up to the next tray above. The composition of the vapor that rises is lighter than the

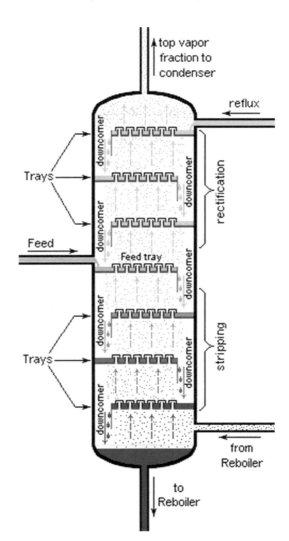

Fig. 4.17 Distillation column schematic

composition of the feed. An even lighter vapor rises from the tray above the feed tray to the tray above that, and so on. Meanwhile some of the liquid on each tray pours over a weir and falls on to the tray below. The composition of the liquid is heavier than that of the vapor. So the liquid on the lower trays consists primarily of the heavier compounds in the feed. The vapor rises continuously in the column and the liquid falls continuously. The figure does not show the reboiler near the bottom of the column where heat is added to the liquid. This heat causes the boiling of the liquids on each tray. The figure also does not show that some side streams with desirable mixtures are usually taken from intermediate trays in the column for further processing.

Refineries rely heavily on distillation, but include other operations. A schematic for a refinery is shown in Fig. 4.18. The figure shows crude oil entering the

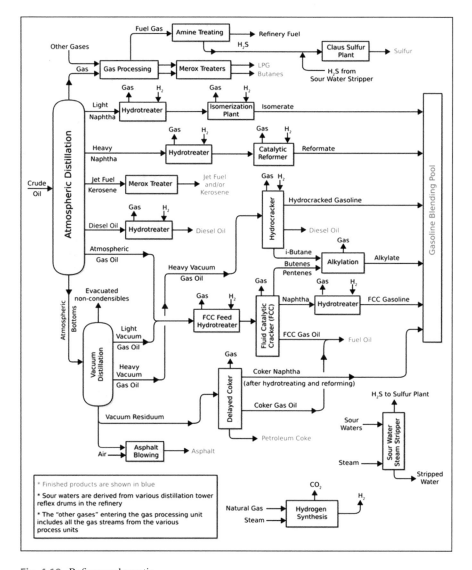

Fig. 4.18 Refinery schematic

atmospheric distillation column at the left. Several liquid side streams are taken from this column: light naphtha, heavy naphtha, jet fuel/kerosene, diesel oil, and atmospheric gas oil. Gas rises from the top of the column. This is a mixture of light hydrocarbons such as propane and butane, which are used for home heating and gas-fired barbecue grills. The light side streams from the upper part of the column are processed further to provide compounds for blending into gasoline. Other side streams provide jet fuel and diesel fuel. The heavy liquid stream from the bottom of the column is sent to a vacuum distillation unit. The boiling points of the compounds in this heavy stream are lower when the pressure is lower, i.e., under a vacuum. This allows further separation of the compounds in that stream at lower temperatures.

The heavy streams undergo various other processes, mainly to convert heavy long-chain hydrocarbons into smaller lighter compounds, which are suitable for blending into various types of gasoline with different octane ratings that you see when you go to a gas station.

The figure also shows processing units that handle hydrogen sulfide (H_2S). H_2S can be processed to recover elemental sulfur, a solid that has various industrial uses including the manufacture of sulfuric acid or cross-linking in rubber. Sulfur is harmful for the catalyst that is used in the fluid catalytic cracker (FCC), so it must not be present in the stream fed to that unit. Some refineries have a hydrodesulfurization unit for removing sulfur.

There are many different types of crude oil worldwide. Refineries tend to be customized for particular types of crude oil, for example favoring production of diesel or gasoline. However, there is some flexibility in the operation of a refinery. It is common for refinery operations to be scheduled based on large constrained optimization models known as linear programs with many variables corresponding to the many streams in the refinery, as well as constraints based on capacity limits of the various units in the refinery. The models accommodate the different prices of the many products supplied by a refinery as well as the cost of inputs used by the refinery.

4.6.4 Worldwide Refining Capacity

Refinery products are indispensable for a modern lifestyle. They can be transported by sea or by pipeline, but it is easier for a country to receive crude oil and refine the product slate within its own borders. There are refineries in at least 78 countries. Table 4.5 lists refinery capacities in various countries in 2019. The USA and China had the largest refinery capacities. This is a reflection of their large demand for refinery products such as gasoline, diesel, and jet fuel.

A list of the world's largest refineries is shown in Table 4.6. The largest refinery, Jamnagar, is privately owned by Reliance Petroleum Limited, and is located in India. It has a capacity of about 1.2 million barrels of crude oil per day. It occupies an area of almost 12 square miles![24] That refinery alone is about a quarter of all the refining capacity in India. The second largest refinery is located in Venezuela, a

---

[24] https://www.hydrocarbons-technology.com/features/feature-top-ten-largest-oil-refineries-world/

Table 4.5 Worldwide refining capacity by country, 2019

Country	Thousand barrels per day[a]
USA	18,974
China	16,199
Russian Federation	6,721
India	5,008
South Korea	3,393
Japan	3,343
Saudi Arabia	2,835
Iran	2,405
Brazil	2,290
Germany	2,085
Canada	2,054
Italy	1,900
Spain	1,586
Mexico	1,558
Singapore	1,514
Netherlands	1,309
United Arab Emirates	1,307
Venezuela	1,303
France	1,245
Thailand	1,235
UK	1,227
Indonesia	1,094
Taiwan	1,083
Other	19,672
Total	101,340

[a]Atmospheric distillation capacity

Table 4.6 World's largest oil refineries

#	Name	Company	Location	Thousand barrels per day
1	Jamnagar Refinery	Reliance Industries Ltd.	Gujarat, India	1,240
2	Paraguana Refinery Complex	PDVSA	Paraguana, Falcon, Venezuela	940
3	Ulsan Refinery	SK Energy	Ulsan, South Korea	850
4	Ruwais Refinery	Abu Dhabi Oil Refining Company	Ruwais, UAE	817
5	Yeosu Refinery	GS Caltex	Yeosu, South Korea	730
6	Onsan Refinery	S-Oil	Ulson, South Korea	670
7		ExxonMobil	Singapore	605
8	Port Arthur Refinery	Motiva Enterprises	Port Arthur, TX, USA	600
9	Baytown Refinery	ExxonMobil	Baytown, TX, USA	561
10	Ras Tanura Refinery	Saudi Aramco	Saudi Arabia	550
11	Garyville Refinery	Marathon Petroleum	Garyville, LA, USA	539
12	Baton Rouge Refinery	ExxonMobil	Baton Rouge, LA, USA	503
13	Abadan Refinery	NIOC	Abadan, Iran	450
14	SAMREF	Aramco Mobil Refinery	Yanbu, Saudi Arabia	405
15	Shell Pernis	Royal Dutch Shell	Rotterdam, The Netherlands	416

country that has large oil reserves. There are large refineries in Southeast Asia, e.g., Korea and Singapore, and in the Middle East. Countries in the Middle East have large oil reserves and are able to capture additional profit from those reserves by processing the crude oil into more valuable products.

4.7   Reserves and Production by Country

Table 4.7 shows the average daily oil production rates for the top ten countries in 2019. It may surprise you to see that the USA produced more oil than Saudi Arabia. Russia was also a large producer. It has significant oil production in Siberia. The table shows that the top ten countries produced about 71% of the world's oil in 2019. So oil production is highly concentrated in those top ten countries.

Table 4.7 Oil production and reserves (2019)

Country	Production		Proved reserves (year-end)		R/P, years
	MBPD	%	Billion barrels	%	
USA	17,045	17.9%	68.9	4.0%	11.1
Saudi Arabia	11,832	12.4%	297.6	17.2%	68.9
Russian Federation	11,540	12.1%	107.2	6.2%	25.5
Canada	5,651	5.9%	169.7	9.8%	82.3
Iraq	4,779	5.0%	145.0	8.4%	83.1
United Arab Emirates	3,998	4.2%	97.8	5.6%	67.0
China	3,836	4.0%	26.2	1.5%	18.7
Iran	3,535	3.7%	155.6	9.0%	120.6
Kuwait	2,996	3.1%	101.5	5.9%	92.8
Brazil	2,877	3.0%	12.7	0.7%	12.1
Other	27,105	28.5%	551.7	31.8%	55.8
Total	95,192	100.0%	1733.9	100.0%	49.9

Crude oil, shale oil, oil sands, condensate, and natural gas liquids

The table also shows the oil reserves in those countries at the end of 2019. The reserves are the amount of oil that is expected to be produced over the life of oil reservoirs that have been discovered.[25] In fact the amount that will be produced is highly uncertain, since the oil is deep underground, and the production rates are uncertain due to technical and economic factors. Low oil prices would induce operators to shut in wells when the revenue is insufficient to cover operating costs. Because of the uncertainty there are several measures of reserves. Proved reserves are the amount of oil that is highly likely to be produced. A probabilistic method can be used to estimate reserves. If so, there is a 90% probability that production will be equal to or in excess of proved reserves. There are other measures of reserves that allow for upside that is likely to occur. These are probable and possible reserves. It is beyond the scope of this book to explain the details of reserve estimation.

---

[25] There is a more speculative category of oil called "resources." This is oil that has not yet been discovered with a well, but is believed to be present based on exploration activity such as seismic.

However, reserve estimation has received a great deal of attention in terms of methods and results.

The named ten countries in the table had 68% of the world's oil reserves as of year-end 2019. Venezuela is reported to have had the largest proved reserves, namely 303.8 billion barrels,[26] which is even larger than the proved reserves in Saudi Arabia! It is possible that the Venezuelan reserves are overstated. Generally, their economy is in poor shape. Specifically, Venezuelan oil production rates have dropped precipitously in recent years as shown in Fig. 4.19. This is an indication that reserves may be falling also. However, the figure shows that reported proved reserves actually tripled from 2007 to 2019. These reported results deserve scrutiny.

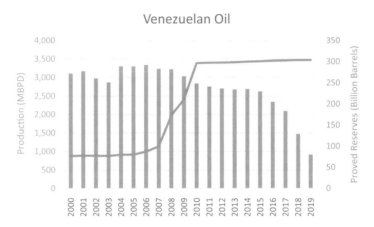

Fig. 4.19   Historical Venezuelan oil production and reserves (year-end)

The rightmost column in Table 4.7 is labeled R/P. This stands for reserve-to-production ratio. It is calculated by dividing the reserves by the production rate and expressing the result in years. The R/P for the USA is 11.1 years. It is important to interpret this number correctly. It does *not* mean that the USA will run out of oil in 11.1 years. There are two reasons that this is the wrong interpretation. First, the production rate in the USA will change over time. The production rates from existing fields typically decline over time as the fields become depleted. This decline may be offset or even reversed by application of the recovery techniques discussed in Sects. 4.4.5–4.4.7. Secondly, new wells will be drilled into existing fields, and new fields will be discovered as the result of exploration activity. So it is best to think of the R/P as the amount of oil inventory. The R/P has been about 10 years in the USA for a long time. There is no need to have more inventory of oil. By way of analogy, a supermarket would need huge warehouses to store 11 years of inventory. It is not necessary to do that. Fortunately, the oil inventory is stored underground in the oil reservoirs.

---

[26] Source [1]

Even with this interpretation, it is notable that some countries have a very high R/P. Many are located in the Middle East: Iran, Iraq, Kuwait, Saudi Arabia, and the United Arab Emirates. These countries have large reserves compared to their production rates. Canada also has a high R/P. It had the fourth largest reserves at the end of 2019. Most of these are tar sands described in Sect. 4.1.4.

4.8 OPEC[27]

The Organization of the Petroleum Exporting Countries (OPEC) was formed at a conference in Baghdad, Iraq, in September 1960. The initial member countries were Iran, Iraq, Kuwait, Saudi Arabia, and Venezuela. Prior to this time, most global oil production was controlled by large oil companies headquartered in the USA, the UK, and the Netherlands, even though much of the production occurred elsewhere. OPEC was formed to enable the producing countries to coordinate their production and gain more control. Several other countries have been members of OPEC, as shown in Table 4.8. Some of these countries are no longer members.

OPEC is discussed more in Sect. 9.2.3.

Table 4.8 Countries that have been OPEC members		
Iran	Qatar	Nigeria
Iraq	Indonesia	Ecuador
Kuwait	Libya	Angola
Saudi Arabia	United Arab Emirates	Gabon
Venezuela	Algeria	Equatorial Guinea
Congo		

4.9 Trade

Table 4.9 summarizes data about global crude oil trade in 2019. The exporting countries are listed on the left column, and the importing countries are listed on the top row. The green cells indicate countries and regions that exported a lot of crude oil in 2019: Russia, Iraq, Saudi Arabia, and West Africa. Perhaps the only surprise in this list of large exporters is West Africa. Nigeria and Angola produced large amounts of crude oil in 2019: 101.4 and 69.1 million tonnes, respectively.[28] Total oil consumption in all of West Africa was only 34.4 million tonnes in that year.[29]

The yellow cells indicate countries and regions that were large importers of crude oil in 2019: the USA, Europe, and China. All of these locations consume significantly more crude oil than they produce. The red cells highlight large specific

---

[27] Source: OPEC.org

[28] Source [1]

[29] Source [1]

Table 4.9 Crude oil trade—2019 (million tonnes)

From \ To	US	Europe	China	India	Japan	Singapore	Other Asia Pacific	Other	Total exports
Canada	189.7	3.7	2.2	0.9	0.0	0.0	0.2	0.4	197.0
Mexico	29.9	10.5	0.5	10.0	0.3	0.0	6.8	0.2	58.1
US	0.0	45.8	6.3	9.1	3.1	1.9	36.0	35.5	137.7
S. & Cent. America	40.0	12.6	67.2	18.7	2.3	0.5	3.9	0.9	146.2
Russia	6.6	153.0	77.7	2.9	7.9	1.4	10.4	26.1	286.1
Other CIS	1.6	68.1	4.2	3.6	0.9	0.4	6.9	5.8	91.5
Iraq	16.5	55.4	51.8	49.2	2.1	2.7	18.9	4.2	200.8
Kuwait	2.2	4.9	22.7	10.6	12.6	2.7	39.8	3.7	99.2
Saudi Arabia	24.9	39.9	83.3	42.6	52.6	6.1	78.5	30.5	358.4
UAE	0.1	0.2	15.3	19.6	42.9	13.2	39.4	8.7	139.4
Other Middle East	0.0	4.3	52.2	10.8	19.3	9.2	23.5	5.9	125.2
North Africa	4.6	59.1	10.7	5.6	0.4	1.4	5.1	6.4	93.4
West Africa	14.1	65.1	77.8	30.2	0.4	2.3	10.0	19.2	219.0
Other	8.1	0.0	35.3	7.8	2.0	7.8	11.1	14.7	86.9
Total imports	338.4	522.5	507.2	221.7	146.9	49.6	290.4	162.2	2,239.0

trade flows. For example, more than half of US crude oil imports came from Canada. Pipelines connect the two countries which facilitates trade. Russia provides a large amount of crude oil to Europe. Saudi Arabia provides large amounts of crude oil to China and other Asian Pacific countries. China also imports large amounts of crude oil from West Africa and Russia. The former is transported in marine tankers, and the latter by pipeline from Russia to China which share a border.

Table 4.10 shows a similar table for trade in petroleum products. The green cells show that both the USA and Russia exported large quantities of products. The yellow cells show that Europe and various Asian Pacific countries were large importers. The red cells show large specific product trade. About 2/3 of Russia's product exports went to Europe. More than half of US product exports went to Mexico and South/Central America. These destinations are relatively close to the USA. Note that Singapore had large product exports from its refineries to various Asian Pacific countries. Note also that crude oil exports from Saudi Arabia and Iraq were much larger than product exports. In contrast, exports of crude and products from the

Table 4.10 Product trade—2019 (million tonnes)

From \ To	Mexico	US	S. & Cent. America	Europe	Middle East	Africa	China	Singapore	Other Asia Pacific	Other	Total exports
Canada	0.6	29.2	1.0	1.7	0.0	0.0	0.9	0.1	0.4	0.9	34.7
US	55.5	0.0	83.0	23.6	2.5	8.6	2.7	2.5	22.1	50.7	251.1
Europe	1.5	24.4	9.6	0.0	13.6	43.9	2.4	16.2	5.1	8.7	125.4
Russia	0.1	18.3	2.8	106.1	3.3	4.8	3.1	9.4	10.1	6.6	164.6
Saudi Arabia	0.0	1.4	0.4	16.5	5.7	10.3	3.5	3.3	6.7	9.6	57.4
UAE	0.0	1.1	1.0	6.7	5.4	11.2	6.6	8.9	23.1	13.1	77.0
Other Middle East	0.0	1.2	0.2	8.1	13.2	5.1	6.9	1.9	14.3	11.3	62.1
China	1.3	0.5	3.7	2.5	1.6	2.8	0.0	15.0	34.1	5.4	66.9
India	0.0	4.6	1.5	10.4	13.5	10.9	1.1	5.3	11.9	1.6	60.7
Singapore	0.1	1.2	1.2	1.7	1.2	3.7	7.1	0.0	58.2	11.6	86.1
Other Asia Pacific	0.6	7.1	0.9	3.8	2.3	5.3	32.8	34.8	0.0	21.7	109.3
Other	1.1	20.7	5.1	28.1	7.2	8.5	11.3	15.1	26.6	22.6	146.5
Total imports	60.9	109.9	110.2	209.2	69.3	115.1	78.4	112.4	212.7	163.8	1,241.9

Source: [1]

UAE were somewhat similar: 139.4 and 77.0 million tons, respectively. There is a large refinery in Abu Dhabi with a capacity of 817 MBPD.[30] Refining capacity in Saudi Arabia is even larger, but its oil production is much larger than that in the UAE (11.8 million barrels per day compared to 4.0 million barrels per day in 2019).[31] Also more products are consumed in Saudi Arabia than in the UAE since its population is about three times the size.

4.10 Summary

This chapter provides information about the oil sector. Crude oil is a mixture of various hydrocarbons. For the most part crude oil is present in reservoirs or shales below the surface of the earth, both onshore and offshore. However, tar sands contain large amounts of heavy oil, known as bitumen. These are mined on the surface. The chapter explains the difference between conventional and unconventional reservoirs, and describes how oil in conventional reservoirs is trapped in porous rocks. It also explains techniques that are used to explore for oil in the subsurface, as well as methods that are used to produce the oil, including fracking. Oil wells have multiple casing strings to protect the subsurface and water aquifers. Primary, secondary, and tertiary techniques for oil production are described. The chapter notes that large amounts of oil are left in the subsurface even after application of these techniques.

Many methods are used to transport crude oil from oil fields to refineries: ships, pipelines, railcars, and trucks. The refineries separate the crude oil into its component parts and process the heavy hydrocarbons into more useful lighter compounds. Refineries produce gasoline, gasoline blending components, diesel fuel, jet fuel, and other products. Blenders produce different types of gasoline using the gasoline blending components.

The chapter lists countries with high oil production rates and high oil reserves. It notes that some countries appear to have only a few years of oil reserves, but this is illusory since new reserves will be discovered. Perhaps surprisingly the USA was the largest oil producer in 2019. Horizontal drilling and fracking have increased US oil production substantially. There is substantial worldwide trade in crude oil and in petroleum products. Tables show the countries and regions that are large exporters and those that are large importers.

4.11 Questions

1. Suppose it takes 2 minutes to fill a 15 gallon gasoline tank in a car. The heat content of gasoline is 115,000 BTU per gallon. What is the rate at which energy is added to the tank in MW?

---

[30] Source: https://en.wikipedia.org/wiki/List_of_oil_refineries#United_Arab_Emirates

[31] Source: [1]

2. There are about 1.1 million active oil and gas wells in the USA. Oil production in the USA was about 17.0 million barrels per day in 2019. What was the average daily oil production rate per well?

3. The US Government has a strategic petroleum reserve (SPR) located underground in Louisiana and Texas in order to cushion against disruptions in oil supply. As of April 2020 it contained 635 million barrels of oil.[32] Average daily US oil production was about 17.0 million barrels per day in 2019.[33] How many days of supply are contained in the SPR, based on the 2019 production rate?

4. Some offshore oil production occurs in locations where the water depth is 10,000 ft. What is the distance in miles from the surface of the water to the seabed with that water depth?

5. The Prudhoe Bay Oil Field in Alaska began commercial oil production in 1977. It originally contained about 25 billion stock tank barrels of oil.[34] 11 billion barrels had been produced through the end of 2005. What fraction of the original oil in place had been recovered by the end of 2005?

6. Oil from Prudhoe Bay and other nearby oil fields flows to the south coast of Alaska through an 800-mile-long 48-in.-diameter pipeline that is part of the Trans-Alaska Pipeline System (TAPS). TAPS also includes a marine terminal at its terminus in Valdez for loading oil tankers. The peak throughput was about two million barrels per day.

   (a) How fast did the oil flow in the pipeline in miles per hour when the production rate was two million barrels per day?

   (b) How many days did it take for oil to travel from the start of the pipeline to Valdez?

   (c) The average daily throughput in 2018 was about 509 thousand barrels per day. How many days did it take oil to traverse the entire pipeline in 2018?

   (d) Unusually for a pipeline, about half of TAPS is aboveground to protect the permafrost soil, since the produced oil is hot. What impact does a lower throughput rate have on the temperature of oil in the pipeline?

7. Production of oil from tar sands has been criticized due to the large amount of energy required to create steam that extracts bitumen from sand. Gas supplies are available near the tar sands, which provide the required energy. Is it wasteful to use natural gas in this way? Why or why not?

8. The reservoir targeted by the Mukluk exploration well mentioned in Sect. 4.2 had the potential to contain at least 1.5 billion barrels. The oil price in 1983, when the well was drilled, was about $29 per barrel. As a rule of thumb, oil in the ground is often valued at ¼ of the market price to account for production cost and the fact that the oil will not be produced immediately, but rather over many years. What was the potential value of oil? Was it a reasonable risk to spend $430 million to drill the exploration well?

---

[32] https://en.wikipedia.org/wiki/Strategic_Petroleum_Reserve_(United_States)

[33] https://www.eia.gov/dnav/pet/pet_crd_crpdn_adc_mbblpd_a.htm

[34] https://en.wikipedia.org/wiki/Prudhoe_Bay_Oil_Field

9. Many people believe that fracking poses large risks to the environment. Do you think this is a reasonable concern?

10. The water cut is the proportion of total liquid production that is water. If the water cut is 80%, what are the relative production rates of water and oil? (Assume that only oil and water are produced.)

11. Table 4.6 shows that the capacity of several large oil refineries is about 500,000 barrels per day. Table 4.1 shows the capacity of various classes of marine oil tanker. How many days of crude oil supply for a 500 MBD refinery are contained in each class of marine tanker?

12. The chemical composition of an alkane is $C_n H_{2n+2}$. Various units in a refinery such as the catalytic reformer and the hydrocracker split large hydrocarbons into smaller ones. In this process a C-C bond is split and a hydrogen atom is added to each of the two smaller hydrocarbons. This requires input of hydrogen. A stylized formula for the chemical reaction in these units is $C_{m+n} H_{2(m+n)+2} + k H_2 = C_n H_{2n+2} + C_m H_{2m+2}$. Solve for the value of k such that there is an equal amount of hydrogen on each side of the equation.

13. Suppose that the inside volume of a cigarette lighter is 0.5 cubic inches. The density of butane vapor at 70°F and 14.5 psia is 0.153 lb. per cubic foot. The density of liquid butane at 70°F and its boiling point is about 36 lb. per cubic foot.

    (a) How many ounces of butane are in a cigarette lighter if it is all vapor?
    (b) How many ounces of butane are in a cigarette lighter if it is all liquid?

14. In 2019 the global company ExxonMobil produced an average of 2386 MBPD of liquids worldwide, refined 3981 MBPD, and sold the product volumes shown below:

	MBPD
Motor gasoline, naphthas	2220
Heating oils, kerosene, diesel oils	1867
Aviation fuels	406
Heavy fuels	270
Lubricants, specialty, and other petroleum products	689
Total	5452

    (a) How does the amount of refining compare with the liquid production? How could the company refine more than it produced?
    (b) How does the amount of product sales compare with the amount of refining? How could the company sell more than it refined?
    (c) What is the amount of each of the products sold as a percentage of the total?

15. Verify the R/P of 11.1 years for the USA shown in Table 4.7.

16. The R/P for the world at the end of 2019 shown in Table 4.7 was about 50 years. Does this mean the world will run out of oil in 50 years? Why or why not?

17. Various types of crude oil are produced in Nigeria. One is Bonny Light with an API gravity of 33.4°.

    (a) What is the density of Bonny Light in kg per m³?
    (b) The text states that Nigeria produced 101.4 million tons of crude oil in 2019. How many barrels per day is that if the average API gravity for all Nigerian crude was 33.4°?

Appendix: Boiling Points and Melting Points

There are three common states of matter: gas (also known as vapor), liquid, and solid. For example ice is solid water, a glass of water contains the liquid form of water, and steam is water in the vapor phase. The stable state of matter depends on its temperature and pressure. There is a specific temperature at which pure compounds[35] change phase. That temperature depends on the pressure. At that specific temperature two phases such as vapor and liquid can coexist. For pure water at atmospheric pressure,[36] the liquid and vapor states can coexist at 212 °F. This temperature is known as the boiling point. If the temperature of a pure compound in the liquid phase is increased above its boiling point, then it becomes a vapor.

There is another temperature at which solid and liquid states can coexist. For pure water at atmospheric pressure, solid and liquid can coexist at 32 °F. This is known as the melting point. If the temperature of the solid form of a pure compound is increased above its melting point, then the matter in the solid state becomes matter in the liquid state. For example, ice melts above 32 °F at atmospheric pressure.

The melting points and boiling points for various straight-chain hydrocarbons[37] are shown in Table 4.11. For example, the straight-chain isomer of octane, the alkane with eight carbon atoms, boils at 258 °F and melts at −71 °F. This means that octane is a liquid between −71 °F and 258 °F. If gasoline were pure octane,[38] then it would remain a liquid at all temperatures that are experienced by drivers, except perhaps in <u>very</u> cold climates. This is good since it would be very inconvenient if gasoline in a tank were to solidify, and very dangerous if it were to vaporize!

---

[35] There are similar concepts for mixtures of compounds, but it is beyond the scope of this book to discuss them.

[36] To be precise, at a pressure of 14.7 pounds per square inch. The boiling point is lower at lower pressures. In particular the boiling point is lower at higher elevations, where the pressure is lower.

[37] As mentioned briefly in Sect. 2.5.1, there are many different forms of alkanes, some of which are straight chains, and some of which are branched. These different forms are known as isomers. The melting points and boiling points of isomers are usually similar, but with small differences.

[38] Gasoline is not pure octane. It is a mixture of many hydrocarbons and may contain other additives.

Table 4.11 Melting points and boiling points of straight-chain alkanes

Number of carbon atoms	Name	Melting point °F	Boiling point °F	State at 70 °F
1	Methane	−297	−259	Gas
2	Ethane	−295	−128	Gas
3	Propane	−306	−44	Gas
4	Butane	−217	31	Gas
5	Pentane	−202	97	Liquid
6	Hexane	−139	156	Liquid
7	Heptane	−131	209	Liquid
8	Octane	−71	258	Liquid
9	Nonane	−63	304	Liquid
10	Decane	−18	345	Liquid
11	Undecane	−15	385	Liquid
12	Dodecane	15	421	Liquid
13	Tridecane	23	453	Liquid
14	Tetradecane	42	487	Liquid
15	Pentadecane	50	516	Liquid
16	Hexadecane	64	549	Liquid
17	Heptadecane	70	576	Liquid/solid
18	Octadecane	84	603	Solid
19	Nonadecane	91	626	Solid
20	Icosane	98	649	Solid
21	Henicosane	105	674	Solid
22	Docosane	108	696	Solid
23	Tricosane	120	716	Solid
24	Tetracosane	126	736	Solid
25	Pentacosane	129	755	Solid
26	Hexacosane	134	774	Solid
27	Heptacosane	139	792	Solid
28	Octacosane	148	809	Solid
29	Nonacosane	147	825	Solid
30	Triacontane	150	841	Solid

At atmospheric pressure

Note that pure forms of heavy hydrocarbons such as icosane are solids at commonly experienced temperatures. Despite this, a liquid-phase mixture of hydrocarbons such as crude oil can contain limited amounts of such heavy hydrocarbons dissolved in the liquid. The solid phase of hydrocarbons can be useful for certain applications, such as solid lighter cubes that are used for charcoal grills.

The melting points and boiling points depend on pressure. Figure 4.20 shows how the boiling point for isobutane varies with pressure. Atmospheric pressure is 101 kPa, shown as blue line in the figure. The boiling point at atmospheric pressure is −12 °C, i.e., 10 °F, which is below freezing.[39] Isobutane is a liquid above the curved line in the figure and a vapor below it. Both liquid and vapor phases can coexist for pressure-temperature combinations on the curved line. The red line in

---

[39] The boiling point for isobutane is lower than the boiling point for straight-chain normal butane shown in Table 4.7.

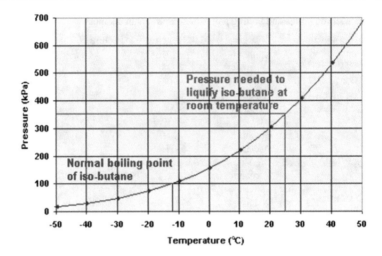

Fig. 4.20   Boiling point for isobutane

the figure shows that isobutane boils at room temperature (25 °C, i.e., 77 °F) if the pressure is about 350 kPa, i.e., 3.5 atmospheres. The contents of a cigarette lighter are at a higher pressure than atmospheric. Lighters contain mostly liquid butane rather than butane vapor at room temperature. This means that the pressure in the lighter is above 350 kPa (about 3.5 times atmospheric pressure). To put this in perspective, a full scuba tank contains air at a much higher pressure of about 13,800 kPa. That is why a scuba tank is made of steel rather than plastic! Similarly, propane tanks for gas barbecue grills contain a mixture of propane and butane. The mixture is at a high enough pressure than the contents are in the liquid phase at typical outdoor temperatures. The contents of the tank vaporize in the barbecue burner where the pressure is lower. It is useful for the contents of barbecue grill tanks and cigarette lighters to be liquid since the density of a liquid is much greater than that of a gas.

Sources

General: [1] BP Statistical Review of World Energy, June 2020

Figures

Number	Source
4.12	https://link.springer.com/article/10.1007/s13202-019-0622-0.
4.14	Medo Hamdani

Number	Source
4.17	H. Padleckas
4.18	https://commons.wikimedia.org/w/index.php?curid=67297997
4.20	https://chemistry.stackexchange.com/questions/86154/why-butane-exist-as-a-liquid-in-a-lighter/86162

Tables

Number	Source
4.1	https://www.planete-energies.com/en/medias/close/transporting-oil-sea and http://alloiltank.com/oil-tanker-ship/
4.2	US Energy Information Agency
4.3	US Energy Information Agency
4.4	US Energy Information Agency
4.5	BP Statistical Review of World Energy, June 2020
4.6	https://www.oilandgasclub.com/worlds-largest-refineries/

Natural Gas

5

5.1 Sources, Exploration, Drilling, and Production

Much of what is described in Chap. 4 for oil also holds true for natural gas. Natural gas occurs in underground reservoirs if there is suitable reservoir rock, a trap with appropriate geometry, a seal to contain the gas, and if there was a source of hydrocarbons as well as a migration path for the hydrocarbons to the reservoir over geological time. As with oil, natural gas that occurs in the subsurface is a mixture of hydrocarbons and sometimes includes impurities such as carbon dioxide, nitrogen, hydrogen sulfide, and/or helium. However, there is no analog such as tar sands for natural gas.[1]

Similar exploration methods are used to find either oil or natural gas. Also, natural gas can be produced either from conventional reservoirs or from tight formations such as shale. In the case of shale, horizontal drilling and fracking are used in the same manner as for oil production.

The previous chapter explained that almost all oil wells produce gas as well as oil. Gas production from oil wells is referred to as associated gas, since it is associated with oil production. Gas wells usually also produce liquid hydrocarbons. Such liquids are often a light form of oil known as condensate. Although most petroleum wells produce both oil and gas they are referred to as oil wells or gas wells depending on the relative amounts of oil and gas.

Similar to oil reservoirs, conventional gas reservoirs may be in communication with aquifers that provide pressure support for the reservoir. However, this is not the case for all gas reservoirs. Also similar to oil reservoirs, natural gas is contained in pore spaces within the reservoir rock. However, unlike oil reservoirs secondary and tertiary recovery methods are not used for gas production. Natural gas has a low

---

[1] However, unlike oil, there are solid gas hydrates in some places at the bottom of the ocean that may become a viable source of natural gas in the future.

© The Author(s), under exclusive license to Springer Nature Switzerland AG 2021
M. Cronshaw, *Energy in Perspective*,
https://doi.org/10.1007/978-3-030-63541-1_5

Global Gas Flaring Reduction Partnership (GGFR)
A large quantity of associated gas is simply flared at oil fields. Oil is a more valuable product than gas. If there is infrastructure for bringing produced oil to the market (such as an oil pipeline) but no such infrastructure for gas, then the gas is burned, since it would be costly to reinject it into the reservoir. Some of the gas may be used to provide energy at the oil field, but often not much. Gas combustion creates carbon dioxide.

The World Bank created the GGFR to reduce this wasteful use of gas.

viscosity and tends not to be trapped in pore spaces by surface tension. So 60–90% of the gas in place can be recovered without secondary or tertiary recovery.

Gas recovery can be enhanced by adding compression at the surface of a well. This lowers the pressure both at the wellhead and at the bottom of the well, which enhances the gas flow rate since it depends on the difference between the pressure in the reservoir and that at the well. Similar to oil wells, gas wells also produce water. Some water vapor is included in any flowing gas, since the gas in the subsurface is in contact with water in the rock pores. If there is pressure support from an aquifer, then water may also flow to a well as a free phase. This liquid water can build up in the tubing string of a well, increasing the bottom-hole pressure in that well due to the hydrostatic head of the water. Artificial lift can be used to produce gas and water from a well that has suffered from such liquid loading.

Water is not the only liquid that affects the performance of gas wells. Just as with pure compounds, the mixture of hydrocarbons in a reservoir can be either in gaseous or in liquid states, or sometimes both states can coexist. This is illustrated in Fig. 5.1. The figure is a map of the state of a particular hydrocarbon mixture depending on its pressure and temperature. The curved line in the figure defines a region within which both liquid and vapor phases coexist. At higher temperatures there is only a vapor phase. At higher pressures there is also only a single phase, referred to as a dense gas.

Fig. 5.1 Natural gas phase diagram

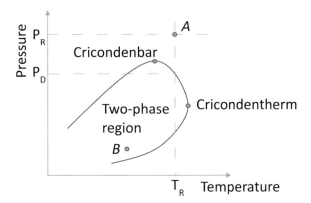

Suppose the reservoir conditions are initially at a high pressure and reservoir temperature T_R shown as point A in the figure. In this case, there is only a single phase in the reservoir. As gas is withdrawn through wells, the pressure falls. The temperature stays constant since the earth continues supplying heat to the reservoir.[2] When the reservoir pressure drops below P_D, two phases (liquid and gas) appear in the reservoir. Both phases are mixtures of hydrocarbons. The liquid phase tends to get trapped in the rock pores due to surface tension. This interferes with fluid flow in the reservoir. Such a reservoir is known as a retrograde condensate reservoir since the liquid that forms is a light oil known as condensate. Such reservoirs require careful management to limit the adverse impact of liquid formation. Sometimes dry gas is injected into the reservoir to maintain the pressure above the level at which liquid forms.

Not all gas reservoirs are retrograde condensate reservoirs. The shape and location of the two-phase region depends on the composition of the hydrocarbon mixture in the reservoir. If it happens that the reservoir temperature is greater than the cricondentherm temperature (see the figure), then the material in the reservoir stays as a single gas phase as the pressure drops, rather than entering the two–phase region.

The discussion above relates to material in the reservoir. During operations the hydrocarbons in the reservoir flow into producing wells. Both the pressure and the temperature fall as this fluid moves up the well. The temperature falls since the fluid loses heat to the earth surrounding the well. The pressure falls due to frictional losses in the well and the reduced hydrostatic head at higher positions in the well. If the separator at the top of the well is at point B in the figure, with a much lower pressure and temperature than in the reservoir, then there will be both liquid and vapor hydrocarbon phases in the separator, just as in an oil well. In addition there will be liquid water due to either water vapor in the hydrocarbon mixture that condenses at lower temperatures or water that flowed from the reservoir to the well as a separate phase. As fluid flows from the reservoir to the separator, it passes from point A in the figure to point B. Even if the fluid in the reservoir remains as a single phase, it will separate into gas and liquid in the well. Both the liquid and gas flow together in the well, a situation known as two-phase flow. Two-phase flow has comingled gas bubbles and liquid, as shown in Fig. 5.2. When there is a relatively large proportion of gas compared to liquid, large bubbles known as slugs, shown on the right side of the figure, tend to form. When fluids in this condition reach the piping in the surface the alternating gas and liquid can cause mechanical problems. Imagine being hit by a series of alternating high-pressure bursts of liquid and gas. Such a series of bursts is challenging for pipes. Equipment known as a slug catcher is often used at the surface to accommodate the alternating liquid and gas.

Two-phase flow also occurs in oil wells. Even if a single oil phase flows in the reservoir, gas comes out of solution as the fluid rises up the well due to falling pressure and temperature. In fact in both gas and oil wells, there may be three phases flowing together: hydrocarbon vapor, hydrocarbon liquid, and liquid water.

---

[2] The rate of pressure decrease in the reservoir is slow so there is no cooling that happens during rapid depressurization of a gas.

Fig. 5.2 Two-phase flow
in a well

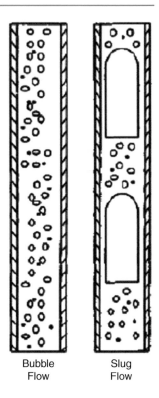

Bubble
Flow

Slug
Flow

5.2   Heat Content

When natural gas is used as fuel, it is the heat content that is important. Heat content
is the amount of energy that is released during combustion of the gas. When any
hydrocarbon is burned in air, the combustion products are carbon dioxide and water
(see Sect. 2.5.5). Since the temperature during combustion is high, the water product is in the form of steam, i.e., water vapor, rather than liquid water. Energy is
released when steam condenses into liquid water. So if the water is condensed, then
more energy is released compared to the energy that is released from combustion
without condensation of the steam. Thus there are two measures for heat content.
The higher heating value (HHV) is the amount of energy released if the steam condenses to liquid and the lower heating value (LHV) is the amount if steam is not
condensed into liquid water. Both the HHV and the LHV are expressed in terms of
BTU per SCF or MMBTU per MSCF. Table 5.1 shows a list of HHVs and LHVs for
natural gas components. Heavier components have higher heating values.

Methane is the main component in natural gas. The heating value of a mixture
can be estimated as the weighted average of the heating values of the separate components based on the composition of the gas. Some produced natural gas contains
inert compounds such as carbon dioxide, nitrogen, and/or helium. These inert compounds have no heating value.

Gas	HHV BTU/SCF	LHV BTU/SCF
Methane	1012	911
Ethane	1783	1631
Propane	2557	2353
Isobutane	3354	3094
n-Butane	3369	3101

Table 5.1 Gas heating values

5.3   Processing

5.3.1   Separation

Gas from several wells is typically collected in an arrangement of surface pipes known as a gathering system. Just as for oil wells, the fluids from the gathering system are sent to a separator such as shown in Fig. 5.3. The gas exits through the top of the separator and the liquids through the bottom. Both the gas and the liquid will contain water. Pipelines that receive the gas do not want large amounts of water since water can freeze during cold weather. The water can be removed with a glycol

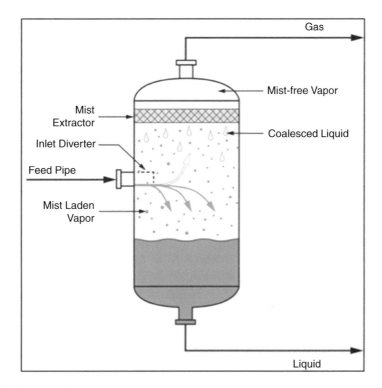

Fig. 5.3 Gas separator

dehydration system, which may be used before the fluids enter the separator. There are also other methods used for dehydration.

The gas and liquid phases may also contain carbon dioxide and/or hydrogen sulfide (H_2S) impurities, which are known as acid gases. Such impurities can be removed by means of amine absorption systems on the surface. The equipment that removes H_2S is known as a sweetening plant, since gas with H_2S is called "sour." Elemental (solid) sulfur can be recovered from the H_2S using the Claus process which requires additional equipment. Alternatively, acid gases that are recovered from an amine system can be injected into subsurface zones that are far removed from aquifers.

The gas from a separator is a mixture of hydrocarbons including methane, ethane, propane, and butanes, sometimes together with the impurities mentioned above. This mixture is referred to as rich gas, in contrast to dry gas which is mostly methane. Dry gas is used by end users as a fuel. Just as in an oil refinery, the compounds in the mixture can be separated by means of distillation. Table 4.11 shows the different boiling points for various gaseous hydrocarbon compounds. You will notice that the boiling points for methane, ethane, and propane are all very low, much below atmospheric temperatures. Thus cold temperatures are used in the distillation process for gas separation. This is known as "cryogenic distillation." Cold temperatures can be realized by dropping the pressure of gas through a valve or through a device known as a turbo expander, which is a turbine.[3] Unlike distillation in oil refineries, cryogenic separation does not involve side streams being taken off the distillation columns. Instead there are separate fractionators known as de-ethanizers, de-propanizers, and de-butanizers. Ethane has a higher heating value than methane that is the main constituent of natural gas. As discussed in Chap. 8, ethane is also a petrochemical feedstock. The extent to which ethane is removed from the natural gas stream depends on its value as a fuel relative to its value as a feedstock. The relative values change frequently depending on supply and demand conditions. So the extent of ethane rejection, i.e., the amount left in the gas stream, can vary on a daily basis.

5.3.2 Hydrocarbon Gas Liquids

Ethane, propane, butane, and natural gasoline are referred to as hydrocarbon gas liquids, even though they may not be liquid at atmospheric conditions. Olefins[4] that are produced in refineries are also classified as hydrocarbon liquids. Figure 5.4 shows that production of hydrocarbon liquids in the USA has increased substantially since 2008.

---

[3] Turbines are also used in the generation of electricity. See Sect. 7.2.

[4] Olefins are hydrocarbons that contain at least one double bond. See Sect. 2.5.2.

U.S. hydrocarbon gas liquids production by type, 2008-2018

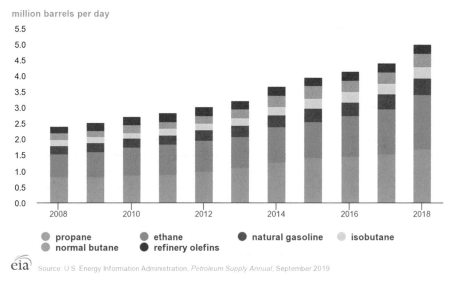

Fig. 5.4 US production of hydrocarbon gas liquids

Production of hydrocarbon gas liquids has increased in line with the increase in US natural gas production, as shown in Fig. 5.5. US natural gas production increased substantially since 2008 due to technological advances in horizontal drilling and fracking.

Ethane has a higher heating value than methane, so it is sometimes included in natural gas for end users. However, it is mainly used as a petrochemical feedstock to produce ethylene. Ethylene can be polymerized to produce polyethylene (see Chap. 8), which is used in plastic bags and as plastic wrap.

You may have noticed the recycling symbols on plastic containers, consisting of small triangles with a single number inside, such as ♲. The numbers correspond to the following polymers:
1. Polyethylene terephthalate
2. High-density polyethylene
3. Polyvinyl chloride
4. Low-density polyethylene
5. Polypropylene
6. Polystyrene
7. Other
Source: https://en.wikipedia.org/wiki/Resin_identification_code#/media/

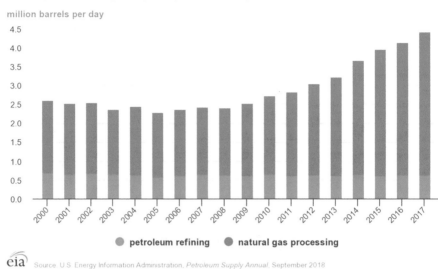

Fig. 5.5   US sources of hydrocarbon gas liquids

Propane has multiple uses. It is a liquid at atmospheric temperatures and at pressures slightly above atmospheric pressure. This means it can be stored as liquid in containers. You have probably used a propane tank for your gas barbecue grill. Propane is widely used for residential and industrial heating in remote locations without supplies of natural gas by pipeline. It is also used as a vehicle fuel in forklifts, lawn mowers, and some other vehicles. Another use is as a petrochemical feedstock to produce propylene. Propylene can be polymerized to produce polypropylene, which has a wide variety of end uses, such as containers, furniture, and even clothing.

There are two isomers of butane: normal butane, which is the straight-chain form, and isobutane, which is the branched form (see Fig. 2.4). Some normal butane is used in cigarette lighters, but most of it is blended into gasoline. It is also used as a petrochemical feedstock to produce butene and butadiene. The former has one double bond and the latter two double bonds. Butadiene can be polymerized to produce synthetic rubber. Isobutane is used to produce alkylates (see Fig. 4.18) that increase the octane of gasoline and control gasoline volatility. Some normal butane is converted to isobutane for this purpose.

Natural gasoline is a mixture of hydrocarbons with five or more carbon atoms. It is used as a gasoline additive. It is also used as a diluent to enable the flow of bitumen from the tar sands in Canada to upgraders (see Sect. 4.1.4).

5.4   Transportation

There are two methods for transporting natural gas: pipelines and various modes for moving liquefied natural gas (LNG). Both are discussed below.

5.4.1   Pipelines

There are 305,000 miles of natural gas transmission pipelines in the USA. That is enough pipe to crisscross the country about 100 times. The gas pipeline network is shown in Fig. 5.6. The figure differentiates between interstate and intrastate pipelines. Interstate pipelines are regulated by the Federal Energy Regulatory Commission (FERC), whereas intrastate pipelines are not. The large number of pipelines in Texas, Louisiana, and Oklahoma is a reflection of the large amount of natural gas that is produced there. There are also a large number of offshore pipelines in the Gulf of Mexico, south of Louisiana, to accommodate the offshore gas production that occurs there.

Gas pipeline diameters range from 6 in to 48 in. Larger diameter pipelines are used for higher gas flow rates and are more costly to build. Gas must be compressed to make it flow through the pipes. Compression also increases the density of the gas, which is important so that it does not occupy too much volume. Usually some of the gas in a pipeline is used as fuel for the compressors. Optimal pipeline design

Map of U.S. interstate and intrastate natural gas pipelines

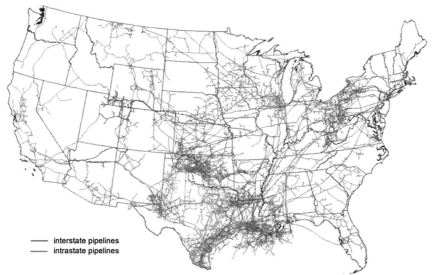

Source: U.S. Energy Information Administration, *About U.S. Natural Gas Pipelines*

Fig. 5.6   Gas pipelines in the USA

involves a trade-off between the diameter of the pipe and the amount of compression that is required. Small-diameter pipe is cheaper to purchase, but requires more compressors at closer spacing for flow of the gas.

Pipelines are typically built out of 40 foot sections of pipe that are welded together during construction of the pipeline and buried at shallow depths below the surface.

Figure 5.6 only shows transmission lines that are used to transport gas from production locations to markets. There are also distribution lines within cities that provide gas to end users. Distribution lines have smaller diameters than transmission lines since their flow rates are lower.

> There is enough natural gas in the Prudhoe Bay Oil field in Alaska to supply 5% of all US needs for about 20 years. In the late 1970s and early 1980s about $600 million was spent on extensive engineering and environmental studies of a proposed Alaskan Natural Gas Transportation System (ANGTS) with a 48-in.-diameter pipeline across Alaska and Canada to bring gas to the Lower 48 states. The USA had experienced severe energy shortages shortly before this time. The idea was abandoned due to its high projected cost. One US senator said that the gas was so remote from markets that it might as well have been on the moon.

5.4.2   Liquefied Natural Gas (LNG)

Natural gas is mostly methane. Gases are not very dense; so a given volume of gas does not weigh very much. One cubic foot of methane at atmospheric conditions (68 °F and 14.5 psi) weighs 0.0412 pounds. If a gas is compressed, then its density increases. One cubic foot of methane at 68 °F and a higher pressure of 1450 psi weighs 4.89 pounds. So if pressure increases by a factor of 100 then the density of methane increases by a factor of 119. It is reasonably straightforward to design a pressure vessel to withstand pressures of 1450 psi. Some vehicles are fueled by compressed natural gas. Their fuel tanks are pressurized vessels that contain natural gas at high pressure. As the fuel is consumed, the pressure in the tank falls (as long as the temperature remains the same).

Instead of increasing its pressure, there is another way to increase the density of natural gas, namely to liquefy it. The boiling point of methane at 14.5 psi is −259 °F. If the temperature of methane falls to below −259 °F then it becomes a liquid. At −259 °F and 14.5 psi, one cubic foot of methane weighs 26.4 pounds, i.e., 640 times as much as at 68 °F and 14.5 psi. So if you want to move a large amount of methane it is good to liquefy it. LNG is mostly methane, but with other compounds. Its density at −259 °F and 14.5 psi is slightly higher than that of liquid methane, namely 27 to 29 pounds per cubic foot.

Transporting natural gas as LNG[5] is becoming increasingly common. LNG is transported in oceangoing tankers (see Fig. 5.7) and in trucks (see Fig. 5.8).

---

[5] Do not confuse LNG (liquefied natural gas) that is mostly methane with NGLs (natural gas liquids). LNG is very cold methane, and NGLs are liquids at atmospheric conditions, typically a mixture of butane, pentane, and heavier compounds.

Fig. 5.7 LNG tanker

Fig. 5.8 Truck hauling LNG

Perhaps as you are reading this you have an insulated cup with coffee by your side, or perhaps you will take a thermos with you to a football game this weekend. These insulated containers are designed to keep the heat from escaping. There is a similar issue with LNG. LNG is really cold! What is the coldest day you have experienced? Maybe −10 °F, or maybe −50 °F. Well LNG at − 259 °F is much colder than that. So heat from the air outside an LNG container flows inwards to the

LNG. Heat travels from a warmer body to a cooler one. Heat flows from your coffee to the air, and from the air to the LNG. So LNG containers must be very well insulated to limit the rate of heat transfer from the outside air.

> There are very few people in northwestern Australia, but there is a lot of natural gas. The offshore Gorgon gas field is located about 12,000 feet below the seafloor. The field produces 2.3 BCFD of gas and 6000 BPD of condensate. Fourteen percent of the gas is CO_2, which is removed at the surface and reinjected into the subsurface. The cleaned methane is converted to LNG and exported to Asia.
>
> Sources: https://australia.chevron.com/-/media/australia/publications/documents/gorgon-co2-injection-project.pdf; https://australia.chevron.com/our-businesses/gorgon-project

5.5   Reserves and Production by Country

Table 5.2 shows the average daily gas production rates for the top ten countries in 2019. It may surprise you to see that the USA was the largest gas producer in 2019. Table 4.7 showed that the USA was also the largest oil producer. Horizontal drilling and fracking have resulted in large increases in the US production of both oil and gas. Russia was the second largest gas producer. It has significant gas production in Siberia. The table shows that the top ten countries produced about 71% of the world's gas in 2019. So, similar to oil, gas production is highly concentrated in the top ten countries.

Table 5.2 Gas production and reserves (2019)

Country	Production		Proved reserves (year-end)		R/P, years
	BCFD		TCF		
USA	89.1	23.1%	455	6.5%	14
Russian Federation	65.7	17.0%	1340	19.1%	56
Iran	23.6	6.1%	1131	16.1%	131
Qatar	17.2	4.5%	872	12.4%	139
China	17.2	4.5%	297	4.2%	47
Canada	16.7	4.3%	70	1.0%	11
Australia	14.8	3.8%	84	1.2%	16
Norway	11.1	2.9%	54	0.8%	13
Saudi Arabia	11.0	2.8%	211	3.0%	53
Algeria	8.3	2.2%	153	2.2%	50
Other	111.1	28.8%	2352	33.5%	58
Total	386.0	100.0%	7019	100.0%	50

The table also shows the proved gas reserves at the end of 2019. The named countries had about 66% of the world's gas reserves. Russia, Iran, and Qatar all had large gas reserves. In fact these were the top three countries for gas reserves. Turkmenistan, in Central Asia, had the fourth largest reserves. In addition the table shows the R/P for each of the countries. The USA, Canada, Australia, and Norway were all large gas producers but they have relatively low R/P ratios. As explained in Sect. 4.7, these low R/P ratios do not mean that those countries will soon run out of gas, since new gas discoveries are likely to be made in each of them. It is notable that both Iran and Qatar have an R/P ratio of more than 130 years, even though they produce large amounts of gas.

Gas has important geopolitical implications. Figure 5.9 shows the locations of Iran, Pakistan, and India. The economies of both Pakistan and India have been growing rapidly. Economic development is strongly correlated with increased energy consumption. Iran borders on Pakistan. For many years there have been discussions about a gas pipeline from Iran to Pakistan and India. Iran's huge gas supplies could provide energy needed by both destination countries. This would benefit all three countries. However, for various reasons, other countries have exerted influence in order to deter construction of such a pipeline. Another proposed

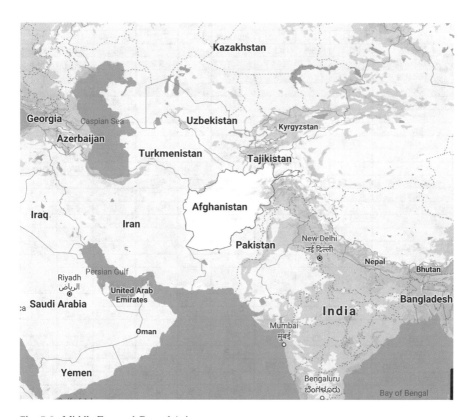

Fig. 5.9 Middle East and Central Asia

pipeline in this region has also been under discussion for years. Turkmenistan's huge gas reserves could also supply gas to Pakistan and India. A pipeline from Turkmenistan to Pakistan and India would cross through Afghanistan. Such a pipeline would benefit all of the countries involved. However instability in the border region between Afghanistan and Pakistan presents challenges for construction and operation of such a pipeline.

5.6   Trade

Both pipelines and LNG are used for international trade of natural gas. Significant volumes of natural gas trade that occurred in 2019 are shown in Table 5.3.

About 2/3 of the trade was via pipeline, and the remainder was by LNG. Consider first the rightmost columns labeled "Net Exports." The green cells in the rightmost column show the top three gas net exporting countries: the Russian Federation, Qatar, and Norway. As shown in Table 5.2 Russia has large gas reserves and was the

Table 5.3   2019 natural gas trade (billion cubic meters)

	Exports			Imports			Net Exports[a]		
	Pipeline	LNG	Total	Pipeline	LNG	Total	Pipeline	LNG	Total
Algeria	26.7	16.6	43.2				26.7	16.6	43.2
Australia		104.7	104.7	5.2		5.2	-5.2	104.7	99.5
Canada	73.2		73.2	24.6	0.5	25.1	48.7	-0.5	48.1
China				47.7	84.8	132.5	-47.7	-84.8	-132.5
France				37.2	22.9	60.1	-37.2	-22.9	-60.1
Germany				109.6		109.6	-109.6		-109.6
India					32.9	32.9		-32.9	-32.9
Iran	16.9		16.9				16.9		16.9
Italy				54.1	13.5	67.5	-54.1	-13.5	-67.5
Japan					105.5	105.5		-105.5	-105.5
Kazakhstan	27.5		27.5	6.9	0.0	6.9	20.6		20.6
Malaysia		35.1	35.1	0.6	3.3	3.9	-0.6	31.8	31.2
Mexico	0.1		0.1	50.8	6.6	57.4	-50.8	-6.6	-57.4
Netherlands	38.2		38.2	40.0	0.0	40.0	-1.8		-1.8
Nigeria		28.8	28.8					28.8	28.8
Norway	109.1	6.6	115.7				109.1	6.6	115.7
Qatar	21.5	107.1	128.6				21.5	107.1	128.6
Russian Federation	217.2	39.4	256.6	26.8		26.8	190.4	39.4	229.8
South Korea					55.6	55.6		-55.6	-55.6
Spain				16.0	21.9	37.9	-16.0	-21.9	-37.9
Trinidad & Tobago		17.0	17.0					17.0	17.0
Turkey				31.3	12.9	44.2	-31.3	-12.9	-44.2
Turkmenistan	31.6		31.6				31.6		31.6
United Kingdom				33.2	18.0	51.3	-33.2	-18.0	-51.3
US	75.4	47.5	122.9	73.3	1.5	74.7	2.1	46.1	48.2
Total World	801.5	485.1	1,286.6	801.5	485.1	1,286.6			

[a] Exports minus imports. Negative numbers indicate net imports

second largest gas producer in 2019. The export columns in Table 5.3 show that most of the Russian exports were by pipeline. These exports mainly go to Europe. Qatar was the second largest net exporter. In contrast to Russia, most gas exports from Qatar were in the form of LNG. Qatar is a small country on the Persian Gulf. It was the fourth largest gas producer in 2019 (see Table 5.2). About 2/3 of its LNG exports went to Asia Pacific countries including South Korea, India, Japan, and China. These destinations are too remote from Qatar for transport by pipeline. Norway was the third largest net exporter. It produces substantial amounts of gas from offshore fields in the North Sea. Almost all of its exports are by pipeline to Europe.

The USA is notably absent from the top three gas net exporters. Table 5.2 shows that it produced more gas than any other country in 2019, and it does export gas both by pipeline and as LNG. However, it also imports a significant amount of gas by pipeline from Canada, so its net exports are not huge.

The red cells in the rightmost column show the top three gas-importing countries: China, Japan, and Germany. Energy demand in China is growing rapidly in line with its rapid economic development. It imports gas as LNG from several countries including Australia, Malaysia, and Qatar. It imports gas by pipeline from Turkmenistan, Kazakhstan, and Uzbekistan (see Fig. 5.9 for a map with the locations of these countries). Japan's electricity supply changed radically after the nuclear accident at Fukushima in 2011. Japan now uses large amounts of natural gas to generate electricity.

The green cells in the "Export" columns of Table 5.3 show countries with large exports. Several of these have been discussed above. Australia produces large amounts of gas offshore in the northwest part of the country. This gas is exported in the form of LNG. The USA exports large amounts of gas both by pipeline and as LNG.

The red cells in the "Import" columns of Table 5.3 show countries with large imports. Large imports by China, Germany, and Japan were discussed above. Italy imports large amounts of gas by pipeline from various countries, including Russia, Algeria, other European countries, and Libya. There are subsea pipelines under the Mediterranean from Algeria (via Tunisia) and Libya to Italy. The USA imports large amounts of gas from Canada. In turn, the USA exports gas to Mexico and also to Canada. It may seem curious that the USA both imports and exports gas to Canada. It is explained by the locations of production. Canada produces a lot of gas in the West of the country, much of which is exported to the USA. On the other hand, Canada uses a lot of gas in the Eastern part of the country. Gas production from the northeastern USA is exported to satisfy that demand in Canada.

5.7   Summary

This chapter explains that there are many similarities between the oil and gas sectors. Most wells produce both oil and gas, as well as water. Natural gas is a mixture of hydrocarbons such as methane, ethane, propane, and butane. Similar to crude oil, distillation is used to separate the different hydrocarbons. Unlike distillation of crude oil, distillation of natural gas occurs at very low temperatures since the boiling points of light hydrocarbons such as methane, ethane, propane, and butane are low.

The energy content of a gas is known as its heating value. The higher heating value (HHV) is the energy released during combustion if the steam formed during combustion is allowed to condense into liquid water. The lower heating value (LHV) is the energy released if the steam is not allowed to condense. Ethane, propane, and butane have higher heating values than methane. However, they are also useful as feedstocks for petrochemicals, so they tend to be removed from produced natural gas rather than being burned.

Natural gas is transported through pipelines and also as liquefied natural gas (LNG). The density of LNG is about 640 times greater than that of natural gas at atmospheric temperature and pressure. So a much greater mass of natural gas can be transported as liquid compared to vapor. LNG is extremely cold, so well-insulated containers are required for its transportation.

The USA was the largest producer of natural gas in 2019. This is due to the success of horizontal drilling and fracking. Russia, Iran, and Qatar have the largest reserves of natural gas. Russia and Qatar export large amounts of natural gas. China, Germany, and Japan import large amounts.

5.8   Questions

1. Suppose that a field produces an average of 2 BCFD. How much gas does it produce in a 365-day year?
2. Suppose that a natural gas stream is 95% methane, 3% ethane, and 2% propane. Estimate the LHV and LHV of the natural gas.
3. 14.1 billion cubic meters of gas were flared in the USA in 2018, while average production was 80.87 BCFD.[6] How much flaring was this as a percentage of all US gas production in that year?
4. Refer to Fig. 5.5.

   (a) From the chart, what was the production quantity of hydrocarbon gas liquids from petroleum refining and from natural gas processing in 2008 and in 2017?
   (b) What was the percentage change in hydrocarbon liquid production from each source during this time period?

5. How many welds are required for a 100-mile-long pipeline assuming that it is built out of 40 foot lengths?

---

[6] http://pubdocs.worldbank.org/en/887251581002821897/Revised-2014-2018-flare-volumes-estimates.pdf

6. One standard cubic foot of gas contains 1.20 moles. A typical molecular weight for natural gas is 16.8 kg/kmol.

   (a) How many kmoles are contained in 1 MSCF?
   (b) How many grams of gas are contained in 1 SCF?
   (c) How many kilograms of gas are contained in 1 MSCF?

7. It is possible to increase the storage of natural gas in gas pipelines by increasing the pressure. Suppose there is a 100-mile-long natural gas pipeline with an inside diameter of 36".

   (a) What is the volume inside the pipeline in cubic meters?
   (b) The density of natural gas at 1400 psi and 60 °F is 4.92 kmol/m^3. How many SCF of natural gas are stored in the pipeline at these conditions?
   (c) The density of natural gas at 1500 psi and 60 °F is 5.31 kmol/m^3. How many SCF of natural gas is stored in the pipeline at these conditions?

8. According to Table 5.2, Russia produced 65.7 BCFD in 2019. According to Table 5.3 it exported 256.6 billion m^3 in that year. What percentage of its production did it export?

9. Using data from Table 5.3, what percentage of gas did Qatar export in the form of LNG?

10. In 2019 Japan consumed 18.67 EJ of energy.[7] What fraction of this consumption was provided by imports of LNG? Assume that 1 billion m^3 of LNG contains 36 petajoules.

11. In 2019 Germany consumed 13.14 EJ of energy.[8] What fraction of this consumption was provided by imports of natural gas? Assume that 1 billion m^3 of natural gas contains 36 PJ.

12. In 2019 China consumed 141.7 EJ of energy.[9] What fraction of this consumption was provided by imports of natural gas? Assume that 1 billion m^3 of LNG contains 36 PJ.

13. In 2019 Australia produced 153.5 billion m^3 of gas.[10] What fraction of this production was exported?

14. In 2019 the USA produced 920.9 billion m^3 of gas.[11] What fraction of this production was exported?

---

[7] Source: [1]
[8] Source: [1]
[9] Source: [1]
[10] Source: [1]
[11] Source: [1]

Sources

General: [1] BP Statistical Review of World Energy, June 2020

Figures

Number	Source
5.7	Pline

Table

Number	Source
5.1	https://www.enggcyclopedia.com/2011/09/heating-values-natural-gas/

Coal

6

6.1   Overview

Coal is an energy source that is used primarily for generating electricity but also for smelting steel. Coal used for electricity generation is pulverized and burned in a furnace to create high-pressure steam in boilers. The high-pressure steam is expanded in a steam turbine that spins an alternator which is the source of the electricity. Coal used in steel manufacture is referred to as metallurgical coal. Metallurgical coal is baked at temperatures up to 1000 °C to drive off volatile constituents. The resulting coke is almost pure carbon. This is combusted with iron ore (iron oxide) in blast furnaces to produce pig iron, which is almost pure iron. Carbon dioxide is a by-product of this process. The formula for the chemical reaction is $2Fe_2O_3 + 3C = 4Fe + 3CO_2$.

There are several types of coal: lignite, subbituminous, bituminous, and anthracite. These different types are referred to as the "rank" of the coal. Anthracite has the highest rank, with a heat content in excess of 12,000 BTU per pound.[1] It contains more than 86% fixed carbon on a dry mineral matter free basis.[2] Lignite, also known as brown coal, has a heat content below 8,300 BTU per pound.[3] Bituminous and subbituminous coal have a heat content above 8,300 BTU per pound and a fixed carbon content below 86%.[4]

Coal contains impurities such as sulfur, mercury, arsenic, and selenium. The sulfur becomes sulfur dioxide, SO_2, when coal is burned. If this is released into the atmosphere with the other combustion gases, it mixes with rainwater to create sulfuric acid. This results in acid rain, which causes environmental damage to forests, rivers, and lakes. The SO_2 can be removed from combustion flue gases by injecting those gases into liquid-filled towers, called scrubbers. The scrubbers produce solid

---

[1] The thresholds for heat content and fixed carbon are approximate.

[2] Source: USGS as reported at https://en.wikipedia.org/wiki/Coal

[3] Ibid

[4] Ibid

© The Author(s), under exclusive license to Springer Nature Switzerland AG 2021
M. Cronshaw, *Energy in Perspective*,
https://doi.org/10.1007/978-3-030-63541-1_6

calcium sulfate, CaSO$_4$, known as gypsum. This is the main component of the building material sheetrock, although most sheetrock is produced from gypsum that is mined as a mineral.

6.2   Reserves, Production, and Trade

As of year-end 2019 the USA was the country with the largest coal reserves, as shown in Table 6.1. The USA had about a quarter of all worldwide proved coal reserves. Russia, Australia, China, and India also had large coal reserves.

Table 6.1   Proved coal reserves by country, YE2019

Country	Anthracite and bituminous	Subbituminous and lignite	Total	% Anthracite and bituminous
USA	219,534	30,003	249,537	88.0%
Russian Federation	71,719	90,447	162,166	44.2%
Australia	72,571	76,508	149,079	48.7%
China	133,467	8,128	141,595	94.3%
India	100,858	5,073	105,931	95.2%
Indonesia	28,163	11,728	39,891	70.6%
Germany		35,900	35,900	0.0%
Ukraine	32,039	2,336	34,375	93.2%
Poland	21,067	5,865	26,932	78.2%
Kazakhstan	25,605		25,605	100.0%
Other	44,144	54,481	98,625	44.8%
Total	749,167	320,469	1,069,636	70.0%

Million tonnes

The table shows that 70% of the world's reserves were the higher rank anthracite and bituminous coal. However, Russia, Australia, Germany, and the rest of the world all have relatively high amounts of the lower rank lignite and subbituminous coal.

As shown in Fig. 6.1, 8.1 billion tons of coal was produced globally in 2019. Thus the reserves at year-end 2019 represent a supply of about 132 years based on 2019 production. Over time, the production rate will change, and new reserves will likely be discovered. Thus the figure of 132 years' worth of supply should not be considered as a definitive prediction of when global coal reserves will be depleted.

Figure 6.1 shows that coal production grew from 1981 to 1989 and fluctuated slightly thereafter until 2001. The average annual increase in coal production in the 31 years from 1981 to 2001 was about 0.8% per year. From 2001 to 2013, coal production increased much more rapidly, about 4.4% per year on average. Coal production declined from 2013 to 2016, and increased from 2016 to 2019.

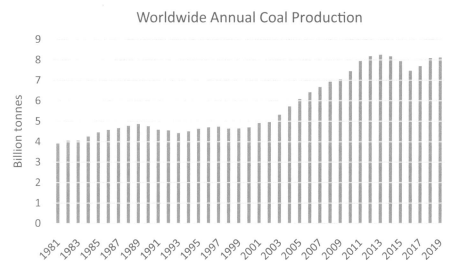

Fig. 6.1   Worldwide coal production

Figure 6.2 shows that the large increase in the amount of global coal production from 2001 was largely driven by increases in China. The economic development that has occurred in China required substantial additional energy use. Coal has been an important factor in the increased energy use. Although coal production in Australia and India is much lower than that in China, production also increased

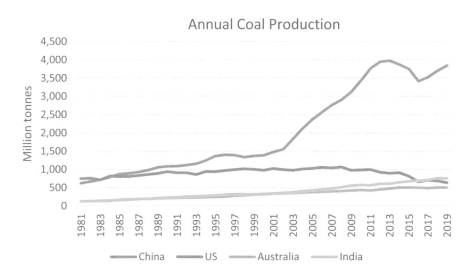

Fig. 6.2   Historical coal production

significantly there; production increased by factors of 3.9 and 5.8, respectively, from 1981 to 2019.[5] Chinese production increased by a factor of 6.2 during the same period, while US production fell by about 14%.[6]

Table 6.2 shows that China produced almost half of the world's coal in 2019 (on an EJ basis). Indonesia was the second largest producer, while the USA, Australia, India, and Russia also had large production. However the total production of all five of these countries combined was less than that of China.

Table 6.2   2019 coal production and consumption (EJ)

Production rank	Country	Production	Consumption	Imports and stock changes
1	China	79.82	81.67	1.86
2	Indonesia	15.05	3.41	−11.64
3	USA	14.30	11.34	−2.96
4	Australia	13.15	1.78	−11.36
5	India	12.73	18.62	5.89
6	Russian Federation	9.20	3.63	−5.57
7	South Africa	6.02	3.81	−2.21
8	Colombia	2.37	0.26	−2.11
9	Kazakhstan	2.08	1.67	−0.41
10	Poland	1.87	1.91	0.04
	Other	10.99	29.75	18.76
	Total	167.58	157.86	−9.72

[1] EJ is equivalent to about 40 tonnes of hard coal or 95 tonnes of lignite / sub-bituminous coal

Table 6.2 also shows consumption by country in 2019. China consumed more coal than it produced, so it must have imported coal, reduced its stocks of coal, or a combination of both. Typically stocks do not change much, so it is reasonable to infer from the table that China was a net importer of coal, as was India. By contrast, Australia, Indonesia, Russia, South Africa, Columbia, the USA, and Kazakhstan were net exporters of coal. The table also shows that the rest of the world (outside of the top ten coal-producing countries) imported large amounts of coal.

Ships are used for most international trade in coal. Figure 6.3 illustrates the trade flows that occurred in 2014. The blue bars on the left of the figure show countries that exported coal, while the green bars on the right indicate importing countries. The width of the bars and of the gray lines between them indicate the volumes of exports, imports, and flows. The diagram shows that Indonesia and Australia were large exporters in 2014, as also shown for 2019 in Table 6.2. The bulk of Indonesian exports went to China, India, Japan, and South Korea, which are relatively close geographically. Australia also exported substantial quantities to those countries. A significant fraction of South African exports went to India.

---

[5] Source: [1]

[6] Ibid

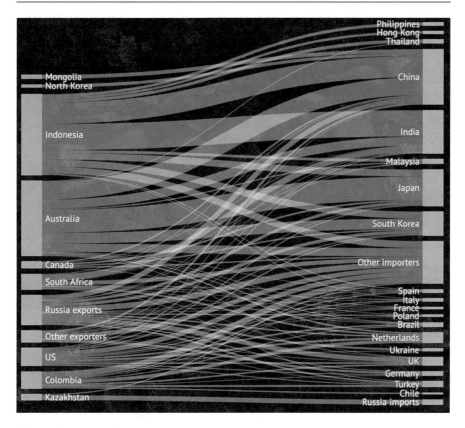

Fig. 6.3 Coal trade (2014)

6.3 Production Methods

Coal is produced both below ground and by surface mining. As shown in Table 6.3, subsurface mining is much more labor intensive than surface mining. Coal production per employee hour in underground mines is only about 1/3 to ½ that in surface mines.

Table 6.3 US coal mining productivity, 2017	Short tons per employee hour	Underground	Surface
	Union	3.91	8.56
	Nonunion	3.76	11.74

Underground coal mining involves cutting coal in tunnels or rooms below the surface, as illustrated in Fig. 6.4. The figure shows that the equipment used in underground mining is approximately on the scale of a human being.

Fig. 6.4 Underground coal mining

By contrast, surface mining uses huge machines as shown in Fig. 6.5. The trucks shown in the figure carry as much as 100 tons of coal. That is more than twice the fully loaded weight of semitrucks that drive on roads! Their fuel tanks hold about 500 gallons, and the diameter of the tires is about 10 ft, more than one and a half times as tall as a human! Based on that image, it is easy to see why surface mining is much less labor intensive than underground mining.

The difference in productivity has led to a rapid increase in US coal surface mining as shown in Fig. 6.6.

There are three main coal-producing regions in the USA that are referred to as basins. The Powder River basin, the Illinois basin, and the Appalachian basin are shown in Fig. 6.7.

Figure 6.8 shows the location of active surface mining locations in 2012, while Fig. 6.9 shows the locations of active underground mining locations. The large number of dots in Kentucky, West Virginia, and Pennsylvania give the impression that most US coal production occurs in those states. However, Table 6.4 shows that coal production in Wyoming dwarfs the production in each of these states. Coal production in Wyoming is predominantly surface mining with its much higher productivity. Figure 6.8 shows only a few surface mining locations in Wyoming, mostly located in the Powder River Basin shown in Fig. 6.7. However, these mines are prolific producers.

Fig. 6.5 Surface coal mining equipment

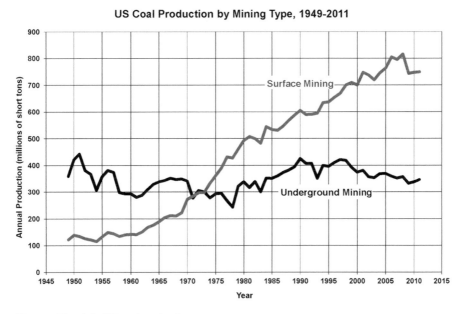

Fig. 6.6 Trends in US coal production

Fig. 6.7   US coal basins

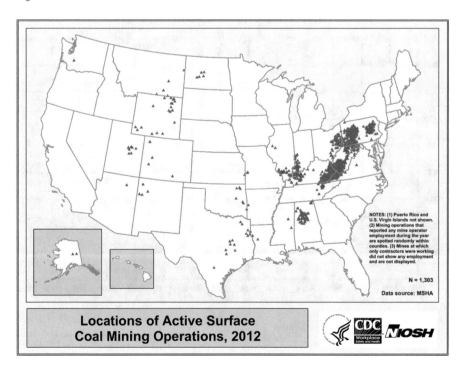

Fig. 6.8   US surface coal mine locations

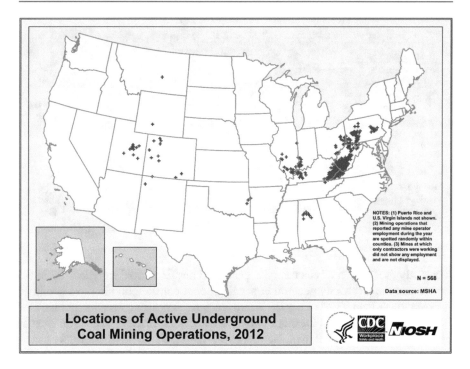

Fig. 6.9 US underground coal mine locations

Table 6.4 US coal production (2018)

State	Coal production ST	%
Wyoming	304.2	40.2%
West Virginia	95.4	12.6%
Pennsylvania	49.9	6.6%
Illinois	49.6	6.6%
Kentucky	39.6	5.2%
Other	217.5	28.8%
Total	756.2	100.0%

6.4   Electricity and Coal

Most coal production is used for generating electricity. As shown in Fig. 3.3 about 90% of US coal production was used for generation of electricity. Table 6.5 lists the amount of electricity generation from coal and the total coal consumption for various countries including the USA. Chapter 7 discusses generation of electricity from various fuels including coal. It explains that only a fraction of the energy in the fuel that is supplied to power plants is converted into electrical energy. Only about 35% of the energy in coal is converted to electrical energy. Given this so-called efficiency

Table 6.5 Electricity production and coal consumption, 2019

Country	Electricity generation from coal, TWh	Coal consumption, EJ	TWh per EJ
China	4854	81.67	59.4
India	1137	18.62	61.1
USA	1054	11.34	92.9
Japan	326	4.91	66.5
South Korea	239	3.44	69.5
South Africa	217	3.81	57.0
Russian Federation	182	3.63	50.2
Indonesia	177	3.41	52.0
Germany	171	2.30	74.3
Australia	149	1.78	83.8
Other	1317	22.94	57.4
Total	9824	157.86	62.2

factor, about 97 TWh of electricity is generated per EJ of coal supplied. Table 6.5
shows that the electricity generation per EJ of coal consumption in most countries
is much lower than 97 TWh per EJ. There are various possible reasons for this. One
is that countries use coal for other purposes besides electricity generation, such as
residential heating or steel manufacture. In this case the amount of electricity generated per EJ of *total* coal consumption would be less than 97 TWh. Another reason
is that the efficiency of the coal-fired power plants may be less than 35%.

As of 2019, about half of global coal-fired power plant capacity was in China.[7]
Figure 6.10 shows the annual additions and retirements to coal-fired power plant

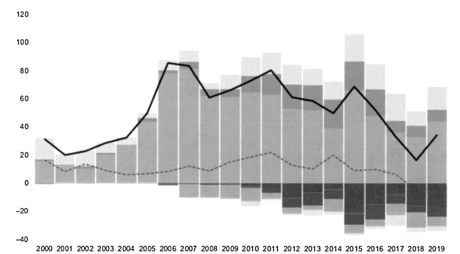

Fig. 6.10 Coal power plant commissioning and retirements (GW)

[7] Global Energy Monitor, Global Coal Plant Tracker, January 2020.

capacity from 2000 to 2019. The bars above zero represent additions and the bars below zero are retirements. It may seem curious that power plants are both added and retired in the same year. But it is normal for old or inefficient plants to be retired in the same year that new plants are added. Another reason for retirements is that coal plants have higher emissions of carbon dioxide than other electricity-generating technologies such as renewables, gas-fired power plants, and nuclear plants (see Sect. 7.5).

The light blue bars in the figure show how China has had the largest additions. The red bars show that the European Union has been retiring coal-fired plants since 2010. The solid black line in the figure shows the net additions, i.e., additions less retirements. Globally, there have been net positive additions in every year since 2000. The dashed line shows net additions for all countries except China. Excluding China, more capacity was retired than added in 2018 and 2019. The rest of the world achieved a net reduction in the capacity of coal-fired power plants in 2018 and 2019 in order to limit emissions of carbon dioxide. However, China is increasing coal-fired electricity generation as its economy expands.

6.5   Coal Transportation

Seventy percent of US coal production is transported by rail.[8] In the USA coal trains (see Fig. 6.11) have about 120 cars with each car carrying about 120 tons.[9] Thus the capacity of a coal train is almost 15,000 tons. This is equivalent to about 375

Fig. 6.11 Coal train

[8] https://www.aar.org/wp-content/uploads/2018/10/AAR-Energy-Issue.pdf
[9] Source: http://www.matts-place.com/trains/coal/coaltrain_basics.htm. Retrieved Jan. 17, 2019.

semitrucks! Trains can operate with a crew of just one or two people. So coal trains offer huge labor savings compared to transport by road. A rapid-discharge bottom dump railcar can empty completely in 5 to 10 seconds.[10] Thus an entire coal train can be unloaded in about 10 to 20 minutes. A large coal-fired power plant consumes about 14,000 tons of coal per day.[11] Such a plant requires about one such coal train delivery each day.

The cost to transport coal is a major factor in the price paid by coal purchasers. This is illustrated in Fig. 6.12. The height of the bars is the delivered price of Powder River Basin coal at power plants, which was about $30 per short ton from 2008 to 2017, although there was some fluctuation. The "market price" shown on the figure is the price of coal at the mine. The delivered price is the sum of the price at the mine and the transportation cost. The transportation cost from 2011 to 2017 was about $20 per short ton. So transportation represented about 2/3 of the delivered cost in that period.[12]

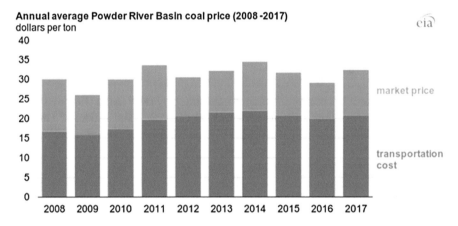

Fig. 6.12 Delivered Price of Powder River Basin coal

6.6   Not all Coal Is Alike

This book avoids most discussions about economics. Energy economics is a fascinating topic, and worthy of study. However, its inclusion would double the length of the book. This subsection is an exception. Table 6.6 shows the prices of coal from various US basins. The prices are reported both in $ per MMBTU and $ per short ton.

You will notice that there is a wide range of prices for coal from the different basins. The price of coal from Central Appalachia was almost five times that from the Powder River Basin on a $ per MMBTU basis and seven times higher on a $/

---

[10] Source: http://www.matts-place.com/trains/coal/car_types-bottomdump.htm

[11] Ibid

[12] The transportation cost was also a significant fraction of the delivered cost before that.

.

The role of coal and trains has changed considerably over time. Nowadays coal trains powered by diesel fuel haul more than 100 cars carrying about 15,000 tons of coal that is mostly used to generate electricity.

In bygone days, steam locomotives such as the one shown above burned coal as fuel. The tender car with the word Jupiter shown behind the locomotive contained the coal that was manually shoveled into the locomotive's boiler to create steam. In the early days of railways electricity had not been invented, and oil (the feedstock for diesel) had not been discovered!

Source: JupiterTrain.jpeg, https://www.nps.gov/gosp/learn/historyculture/everlasting-steam-the-story-of-jupiter-and-no-119.htm

Table 6.6 US coal spot prices (Week of 1/11/19)

Location	Heat content BTU/lb.	Sulfur lb. SO$_2$ per MMBTU	\$ per MMBTU	\$ per short ton
Central Appalachia	12,500	1.2	\$3.36	\$83.90
Northern Appalachia	13,000	<3.0	\$2.53	\$65.70
Illinois Basin	11,800	5.0	\$1.65	\$38.95
Powder River Basin	8,800	0.8	\$0.68	\$11.90
Uinta Basin	11,700	0.8	\$1.72	\$40.25

short ton basis. This is such a big range of prices that there must be something really different about the coals from the various basins.

Indeed, the coals differ in three ways: location, heat content, and sulfur content. As explained in Sect. 6.5 the cost to transport coal is a very important component of its delivered cost. The transportation cost explains why location is so important. Coal buyers are interested in the price of coal at their location. Buyers will not be willing to pay more for coal of the same quality no matter where it comes from.

Transportation cost increases with distance. So the netback price[13] of coal at a mine decreases with distance to market. That is, the price of coal at a mine that is far from a market tends to be lower, other things being equal. However, usually other things are not equal.

The table shows that the heat contents range from a low of 8,800 BTU per pound in the Powder River Basin to a high of 13,000 BTU in Northern Appalachia. The difference in heat content is reflective of the different ranks of coal. Coal from the Powder River Basin is subbituminous and that from Appalachia is bituminous. The heat content is important since about 90% of US coal is used for power generation, and the chemical energy contained in the coal is its source of value for electric power generation. Electricity is discussed in Chap. 7.

Table 6.6 also shows that the sulfur content of different US coals varies considerably. The sulfur content is reported in terms of pounds of SO_2 per MMBTU. Coal does not contain SO_2, which is a gas. However, coal does contain sulfur. This sulfur is converted into SO_2 during combustion of the coal. If SO_2 is emitted into the atmosphere, it mixes with water and then falls to the earth as acid rain. This acid rain kills trees as well as acidifies lakes, rivers, and streams. SO_2 emissions from US power plants are regulated due to these adverse environmental impacts.

Notice that the sulfur content increases and the price decreases moving down the first three coals in the table. Central and Northern Appalachia are nearby each other and the heat content of the two coals is similar. So the difference in sulfur content must explain a large part of the difference in price. The Illinois basin coal has the highest sulfur content, but it is in quite a different location and has a lower heat content. Each of these differences contributes to the difference in price for Illinois basin coal relative to other coals.

The difference in the prices of the Appalachian coals can be explained by the US regulation of sulfur dioxide emissions from coal-fired power plants. The USA implemented an innovative system of regulation with the 1990 Clean Air Act Amendments. This regulation is widely regarded as having been successful. Power plants are required to surrender allowances in proportion to their emissions of SO_2. A finite number of allowances are issued to utilities each year. They are tradeable. Plants can choose to use low sulfur coal and/or to install scrubbers to remove SO_2 from their flue gas emissions. Plants that find it cost effective to undertake such abatement measures will do so, and sell their allowances to offset their cost of abatement. On the other hand, firms that find it costly to abate can buy allowances from sellers. This tradeable allowance scheme achieves two goals: (1) the total emissions are capped by the total number of allowances that are issued,[14] and (2) firms with low abatement cost do the abatement, whereas those that would have high abatement costs are not required to incur these high costs.[15]

---

[13] Netback price is the price of a product minus the cost to transport the product from source to customer.

[14] Allowances can be banked and used in later years. So emissions in some future year could exceed the allowances issued in that year due to use of previously banked allowances.

[15] The decentralized nature of allowance trading means that a central regulator does not require knowledge of the abatement costs for different plants. The allowance market automatically sorts this out.

The bottom line of this allowance program is that US power plants incur costs associated with SO_2 emissions, due to the cost of either physical abatement or buying allowances. Because of this, low-sulfur coal is more valuable to a power plant than high-sulfur coal. That is, power plants are willing to pay a higher price for low-sulfur coal. This is a major reason for the higher price of coal from Central Appalachia compared to coal with more sulfur from Northern Appalachia. The impacts of location and heat content on coal prices are discussed further in an appendix to this chapter.

6.7 Summary

This chapter explains the two methods for coal mining: underground and surface. Both are in common use. There are various types of coal, with different heat contents. Lignite has the lowest heat content. Subbituminous coal, bituminous coal, and anthracite have higher heat contents with anthracite being the highest.

The USA, Russia, Australia, China, and India all have large reserves of coal. The USA, Russia, and Australia all export coal, as do Indonesia, South Africa, and Colombia. Despite their large reserves, China and India import coal to supplement their domestic production.

Most coal is used in coal-fired power plants to generate electricity. Since most coal-fired plants are remote from coal mines, transportation adds a significant amount to the delivered cost of coal. Coal is transported by rail over land, and by sea on ships.

Some coal contains sulfur and other impurities. The sulfur is converted to sulfur dioxide when the coal is burned. If the sulfur dioxide is emitted to the atmosphere with the other combustion products, then it mixes with water in the atmosphere to form sulfuric acid and falls to the earth as acid rain causing environmental damage. Power plants can install scrubbers to remove sulfur dioxide from the flue gases.

6.8 Questions

1. This question is about US coal reserves and production.

   (a) Use the data in Table 6.1 to determine what percentage of the world's proved coal reserves were located in the USA as of year-end 2019.
   (b) Use the data in Table 6.2 to determine the US coal production in 2019 as a percentage of the world total.

2. This question is about Chinese coal reserves and production.

   (a) Use the data in Table 6.1 to determine what percentage of the world's proved coal reserves were located in China as of year-end 2019.
   (b) Use the data in Table 6.2 to determine the Chinese coal production in 2019 as a percentage of the world total.

3. Figure 6.6 shows trends in US coal production.

   (a) Read the graph to determine the amounts of coal production by surface min-
       ing and by underground mining in 1949 (the leftmost points on the curves).
   (b) What percentage of the coal production was by underground mining in 1949?
   (c) Answer the same questions for 2011 (the rightmost points on the curves).

4. India produced 223.3 million tonnes (4.45 EJ) of coal in 1990, and 756.4 million
   tonnes (12.73 EJ) in 2019.[16]

   (a) What was the average heat content (GJ per tonne) of the coal in 1990?
   (b) What was the average heat content (GJ per tonne) of the coal in 2019?
   (c) What do you infer about the changes in the types of coal produced in India
       over that period?

5. The text states that about 90% of US coal production is used for the generation
   of electricity.

   (a) Table 6.2 shows that the USA consumed 11.34 EJ of coal in 2019. What
       amount of this coal was used for electricity generation?
   (b) Table 6.5 shows that 1054 TWh of electricity was produced from coal in
       2019. What percentage of the energy that was supplied in the coal is pro-
       duced in the electricity?

6. Verify the statement in Sect. 6.4 of the text that 97 TWh of electricity would be
   generated per EJ of coal feed to a coal-fired power plant if its efficiency is 35%.
   (Hint: Use a conversion factor from Table 2.14.)

7. Table 6.5 shows that the USA produced 92.9 TWh of electricity per total EJ of
   coal consumption in 2019. The text states that a coal-fired power plant would
   produce 97 TWh of electricity per EJ. Provide a possible explanation for why the
   actual number in the USA is lower than 97 TWh per EJ.

8. Table 6.5 shows that China consumed 81.67 EJ of coal in 2019 and produced
   59.4 TWh of electricity per EJ of coal consumption. How much electricity would
   it produce per total EJ of coal consumption if it used 70% of its coal for electricity generation and the average efficiency of its coal-fired power plants was 30%?
   Compare your answer with the 4,854 TWh of generation from coal shown in
   Table 6.5.

9. Suppose that the efficiency of a coal-fired 500 MW power plant is 35%, and the
   heating value of the coal fuel is 12,000 BTU per pound. Assume that the power
   plant operates at full load all of the time.

   (a) How many short tons of coal does the plant burn per day?
   (b) How many coal railcars per day are needed to deliver that amount of coal, if
       each car carries 120 short tons?

---

[16] Source: [1]

Appendix: Coal Price Variation

There is a wide variation in the coal prices shown in Table 6.6. This indicates that there must be important differences in the types of coal listed in the table. This appendix explores those differences, and provides a plausible explanation for them. The methods used here illustrate how the basic understanding of energy that is described in this book can explain energy data.

Heat Content

As a first point, note that the table reports prices both in $ per short ton and in $ per MMBTU. It is easy to convert between the two. For example, the table shows that Central Appalachia has a heat content of 12,500 BTU per pound and had a price of $83.90 per short ton in the week ending January 11, 2019. 12,500 BTU is 0.0125 MMBTU. A short ton contains 2000 pounds, so the price is

$$\frac{\$83.90}{ST} \times \frac{1\,ST}{2000\,lb} = \$0.042 \quad \text{per pound.}$$

To determine the price per MMBTU, one uses the heat content of the coal:

$$\frac{\$0.042}{lb} \times \frac{1\,lb}{0.0125\,MMBTU} = \$3.36\,per\,MMBTU,$$

which agrees with the value shown in the table.

Powder River Basin vs. Uinta Basin

Consider the price difference between Powder River and Uinta Basin coals. The Powder River Basin is located in Wyoming whereas the Uinta Basin straddles Colorado and Utah (see Fig. 6.13).

Table 6.7 shows the amounts of coal that were produced and consumed in those three states in 2017 and 2018. The table also shows numbers for "net consumption" which is the amount of consumption minus the amount of production. If net consumption is positive, then more coal was consumed than produced in that state. If it is negative, then more coal was produced than consumed.

Note that Colorado consumed 10–12% more than it produced, whereas Utah consumed 7–10% less than it produced. The consumption in the two states together was about the same as the amount of production. If you had data about the destination of produced coal, then you would know about coal transportation across state lines. The rail companies that transport coal have such data, but might keep it private. Since the net consumption for the two states together is close to zero, it is reasonable to assume that most of the coal produced in those two states is consumed within them.[17] If so, then Uinta Basin coal is not transported for long distances.

---

[17] There may actually be exports and imports with other states, but they are probably small.

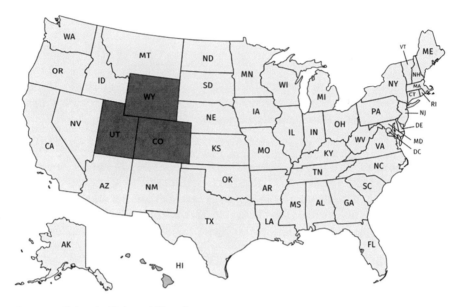

Fig. 6.13 Colorado, Utah, and Wyoming

Table 6.7 Coal production and consumption in selected states

	Thousand ST							
	Production		Consumption		Net consumption		Net consumption %	
	2018	2017	2018	2017	2018	2017	2018	2017
Colorado	14,026	15,047	15,445	16,923	1,419	1,876	10%	12%
Utah	13,619	14,326	12,710	12,923	−909	−1,403	−7%	−10%
Total	27,645	29,373	28,155	29,846	510	473	2%	2%
Wyoming	304,188	316,454	25,969	26,302	−278,219	−290,152	−91%	−92%

Contrast that with Wyoming. The table shows that more than 90% of Wyoming's coal production was exported out of the state, since net consumption was −91% and −92%. Most Powder River Basin coal is transported to the Midwestern USA and beyond. This large export percentage is consistent with Table 6.4, which shows the large percentage of US coal that is produced in Wyoming. The low sulfur content of Powder River Basin coal is attractive to coal-fired power plants, whose emissions of sulfur dioxide are regulated.

Table 6.6 shows that the sulfur content of Powder River Basin coal is the same as that of Uinta Basin coal. So sulfur is not the reason for the different prices. The different heat contents[18] shown in the table explain part of the reason, but the major difference is due to different coal transportation costs. Figure 6.12 shows that the

---

[18] The difference in heat content is a consequence of different coal ranks. Uinta Basin coal is bituminous, whereas Powder River Basin coal is subbituminous.

transportation cost for Powder River Basin coal was about $20 per short ton from 2011 to 2017. Let us assume that it was $20 per short ton in 2018 and 2019, the years shown in Table 6.6.

The purchasers of coal pay both for the coal and for its transportation. According to Table 6.6, Powder River Basin coal producers received $11.90 per short ton for their coal in the week ending January 11, 2019, so purchasers would have paid $11.90 + $20 = $31.90 per short ton for delivered coal. Since the heat content of that coal is 8,800 BTU or 0.0088 MMBTU per pound, the equivalent price is

$$\frac{\$31.90}{ST} \times \frac{1\,ST}{2000\,lb} \times \frac{1\,lb}{0.0088\,MMBTU} = \$1.81\,per\,MMBTU.$$

Let us make another assumption, namely that coal-fired power plants pay the same price for coal on a BTU basis wherever the plants are located. This is a reasonable assumption if the sales price of the electrical power is similar and if the economics of their plants are also similar.

With this assumption the delivered price per short ton paid for Uinta Basin coal is also $1.81 per MMBTU. Since its heat content is 11,700 BTU or 0.0117 MMBTU per pound, the price per short ton is

$$\frac{\$1.81}{MMBTU} \times \frac{0.0117\,MMBTU}{1\,lb} \times \frac{2000\,lb}{1\,ST} = \$42.41\,per\,ST.$$

Table 6.6 shows that the mine-mouth or netback price of Uinta Basin coal was $40.25 per ST, so the inferred transportation cost for that coal was $42.41 − $40.25 = $2.16 per short ton. This is about 1/10 the transportation cost for Powder River Basin coal, which seems entirely reasonable based on the much shorter haul distances for Uinta Basin coal compared to Powder River Basin coal.

Sources

General: 1. BP Statistical Review of World Energy, June 2020

Figures

Number	Source
6.2	[1]
6.3	https://www.carbonbrief.org/mapped-the-global-coal-trade
6.4	https://goo.gl/images/UjKAYJ
6.6	By Plazak—own work, CC BY-SA 3.0, https://commons.wikimedia.org/w/index.php?curid=27584769
6.7	http://phinix.net/blog/tag/coal-production-2/, Retrieved Jan. 18, 2019
6.8	US Dept. of Health, Centers for Disease Control and Prevention
6.9	US Dept. of Health, Centers for Disease Control and Prevention
6.10	Global Energy Monitor, Global Coal Plant Tracker, January 2020, BoomAndBust_2020_English.pdf

Number	Source
6.11	https://www.quora.com/ Which-will-produce-more-electricity-115-rail-cars-of-coal-each-day-for-30-years-or-installing-a-train-load-of-solar-panels-once
6.12	https://www.eia.gov/todayinenergy/detail.php?id=41053
6.13	https://bigthink.com/strange-maps/colorado-is-not-a-rectangle?rebelltitem=6#rebelltitem6

Tables

Number	Source
6.2	[1]
6.3	US Energy Information Agency
6.4	US Energy Information Agency
6.6	https://www.eia.gov/coal/markets/#tabs-prices-2, Retrieved Jan. 14, 2019
6.7	https://www.eia.gov/coal/data.php#production

Electricity

7

7.1   Electrical Systems

7.1.1   Large Scale

Electricity is not a primary energy source, unlike oil, gas, and coal. Instead electricity is generated using a primary energy source and delivered to customers. The conversion of fuel to electricity occurs because electricity is a very convenient form of energy. It does not need to be stored and is available with the flick of a switch. Batteries can store electrical energy, but it is technically difficult to store large amounts, although that is changing.[1] There are two types of electrical current: alternating current (AC) and direct current (DC). With AC the voltage and current fluctuate over time in the shape of a sine wave. In North American electrical systems, the frequency of AC is 60 Hz, i.e., 60 cycles per second. In Europe, the frequency is 50 Hz.[2] By contrast, the voltage in DC systems is constant and does not fluctuate over time. Household batteries such as AAA, AA, C, or D sizes deliver DC. There are some specialized uses of high-voltage DC systems, but they are not discussed in this book.

Large-scale electrical systems have three aspects, as shown in Fig. 7.1: generation, transmission to load centers (such as towns), and distribution to end users. Some large industrial customers receive electricity directly from the transmission system, bypassing the distribution network, as shown in the figure. The whole system is often referred to as the electrical grid.

In addition to the system in the figure, some customers have solar panels (see Fig. 7.2) and/or wind turbines so they generate electricity on their premises. Solar panels are referred to as photovoltaic, since they transform light into electricity. Such generation can be used for a customer's own consumption and/or sent into the

---

[1] Storage of electricity is discussed below.

[2] For reasons that are beyond the scope of this book, AC systems use three-phase power, with three separate alternating voltages.

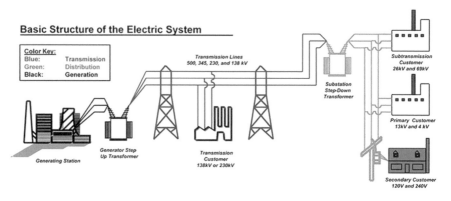

Fig. 7.1 Electrical system

Fig. 7.2 Home solar panels

grid. Such distributed facilities do not require transmission. They also do not need distribution, unless they send power to the grid. But many such systems are connected to the grid so that the residence can use grid power if necessary, for example if the sun is not shining or the wind is not blowing.

7.1.2   Microgrids

It may not be economical to connect remote locations to large-scale electrical systems. Microgrids are an alternative solution for providing electrical energy to remote locations. They are a decentralized group of electricity sources[3] (such as photovol-

---

[3] These sources are described below.

taic, wind, and/or diesel) and loads. They often include batteries for energy storage. Some less remote microgrids are also connected electrical grids.

7.2   Generation

Electricity is generated in several different ways: using fossil fuels (coal, natural gas, liquid fuels such as diesel), nuclear, hydroelectric, solar, and wind. Figure 7.3 shows the shares of fuel types that were used to generate electricity in the USA in 2019. Natural gas provided about 39%. It was the largest fuel type. Coal was the second largest, having provided 24%. Nuclear energy provided about 19%; renewable energy, i.e., solar and wind, provided about 11%, while hydroelectric, i.e., dams, provided about 6%.

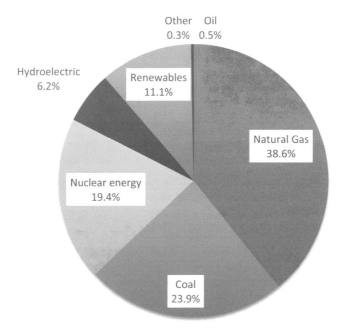

Fig. 7.3 Generation of electricity by type of fuel (USA, 2019)

The mix of fuel types used for electricity generation is quite different in other countries. For example, Fig. 7.4 shows that about two-thirds of the electricity in China was generated by coal in 2019. Natural gas and nuclear power provided much smaller shares than in the USA, while hydroelectricity provided a much larger share. The share of renewables was almost the same as in the USA.

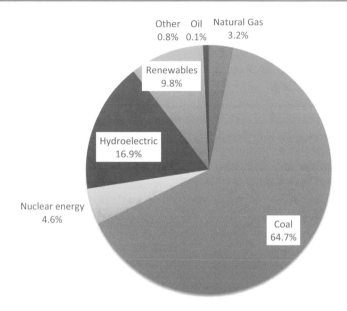

Fig. 7.4 Generation of electricity by type of fuel (China, 2019)

7.2.1 Coal-Fired Power Plants

Coal-fired power plants transform the chemical energy in coal into electrical energy. A schematic of a coal-fired power plant is shown in Fig. 7.5. Coal is pulverized, and then burned with air in a boiler. The heat released during combustion is used to convert high-pressure liquid water into high-pressure steam, which is the vapor

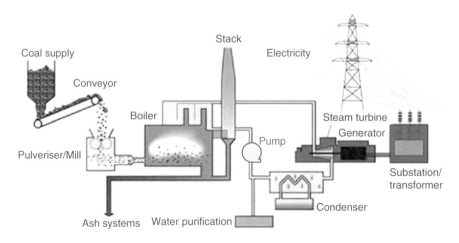

Fig. 7.5 Coal-fired power plant schematic

phase of water. The combustion gases from the boiler are vented to the atmosphere in a chimney stack. The high-pressure steam is expanded through a steam turbine that is connected to a generator (also known as an alternator). The expanding steam spins both the turbine and the generator. The generator contains coils of wire on a rotor that turns in a magnetic field. Turning in a magnetic field generates electricity. The rotation of the alternator is what generates AC power. The electricity is sent to a transformer where the voltage is increased to high levels and sent to a transmission line. Losses in transmission lines are lower when the voltage is higher (see Sect. 7.3).

As the steam expands in the turbine, it cools and partially returns to its liquid state. The water that leaves the turbine is further cooled in a condenser and returned entirely to its liquid state. There are various types of condenser. Some use water from a nearby lake, river, or ocean. Others use cooling towers, as shown in Fig. 7.6. Hot water entering a cooling tower is used to vaporize separate cooling water that escapes from the top of the tower as shown in the figure. What looks like pollution escaping from the tower is simply water vapor, like a cloud!

Fig. 7.6 Cooling towers

Once the water/steam mixture that enters the cooling tower has been cooled and returns entirely to its liquid state, it is pumped up to a high pressure and the water/steam cycle repeats itself. It is easier to increase the pressure of water in its liquid form than to do so when it is vapor. This is why cooling is important. The high-pressure steam does work by driving the turbine. This work is converted into electrical energy in the alternator.

The coal burned in power plants consists mostly of carbon and hydrogen. However there can also be impurities including sulfur, mercury, and noncombustible materials. The combustion products of carbon and hydrogen are carbon dioxide and water. Due to the high combustion temperature, the water is in the form of steam when it leaves the chimney stack. The white plume that you see coming out of chimney stacks appears to be pollution, but it is water. The carbon dioxide that also escapes up the stack is not visible. However coal-fired power plants are one of the main sources of carbon dioxide emissions into the atmosphere, the same carbon dioxide that has been associated with global climate change.

During combustion, sulfur contained in the coal is converted to sulfur dioxide. If this is emitted into the atmosphere it reacts with water to form sulfuric acid. This acid then falls to the earth as acid rain which is associated with degradation of plant life and acidification of rivers and lakes. The sulfur dioxide can be removed from combustion gases by passing them through flue gas desulfurization units, also known as scrubbers. There are both capital and operating costs associated with scrubbers. Also, scrubbers consume energy, which reduces the net amount of energy available from the power plant.

Coal-fired power plants also emit particulate matter, nitrogen oxides, and mercury, all of which can be harmful to human health. Since the air used in combustion contains about 78% nitrogen, there is little that can be done to control the emission of nitrogen oxides. However, careful control of combustion can have some beneficial effect. Additional steps are taken to reduce the emissions of particulates and mercury. There are special designs for clean coal technologies that control emissions from coal-fired power plants. These types of plants have not been deployed on a large scale.

In 2017 there were 789 coal generators in the USA with a total nameplate capacity of 279.2 GW.[4] Thus the average generator size was about 350 MW. However there is considerable variation in the size of coal-fired power plants. The Robert Scherer power plant in Georgia has the largest nameplate capacity: 3520 MW.[5] This plant has four units rated at 880 MW each; it has two stacks that are 1001 ft tall.[6] The plant receives 2–5 unit trains per day with coal from the Powder River Basin in Wyoming.[7] It emits over 20 million tons of greenhouse gases per year.[8] One of the four units is planned to be shut down by 2022.[9]

---

[4] https://www.eia.gov/electricity/annual/html/epa_04_03.html

[5] https://en.wikipedia.org/wiki/List_of_the_largest_coal_power_stations_in_the_United_States

[6] https://en.wikipedia.org/wiki/Plant_Scherer

[7] Ibid

[8] Ibid

[9] Ibid

7.2.2   Gas-Fired Power Plants

There are two technologies for gas-fired power plants: rotating gas turbines and reciprocating engines with pistons similar to those in car engines. The main component of a turbine gas-fired power plant is similar to a jet engine, but with one important difference. A jet engine burns fuel so as to thrust hot exhaust gases (and bypass air) backwards. According to Newton's third law, there is an equal and opposite reaction that propels the jet forwards. In contrast, the gas turbine in a power plant does not create thrust with exhaust gases. Instead it is connected to an alternator similar to that in a coal-fired plant. The alternator generates electricity. Similar to a jet engine, the gas turbine also emits hot exhaust gases. In a combined cycle gas-fired power plant, these hot gases are used to generate steam in a heat recovery steam generator (HRSG). This steam is passed to a steam turbine that generates additional electricity, similar to the steam turbine in a coal-fired plant. The schematic of a combined cycle gas-fired power plant is shown in Fig. 7.7. The figure shows natural gas entering the combustion chamber (shown in green in the middle of the figure) where it is burned with air. The hot exhaust gas from the gas turbine enters the boiler on the right of the figure. The boiler creates steam that is fed to the steam turbine. Both the steam turbine and the gas turbine are connected to alternators.

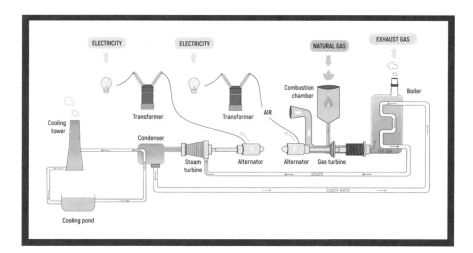

Fig. 7.7   Gas-fired power plant schematic

A combined cycle gas-fired power plant may have one or more steam turbines. Some gas-fired power plants do not have any steam turbines. Such plants are referred to as simple cycle power plants.[10] Combined cycle power plants are able to convert

---

[10] There are also some gas-fired power plants that only have steam turbines, without any combustion turbines. They use natural gas rather than coal to generate steam.

more of the natural gas fuel's chemical energy into electricity, compared to simple cycle plants. The effectiveness of the energy conversion is referred to as efficiency, which is discussed below.

Some gas-fired power plants can be fueled either with natural gas or with liquid fuel such as diesel. Typically neither of these fuels contains sulfur or other impurities. So the emissions from a gas-fired power plant are almost entirely carbon dioxide and water, and do not include the noxious by-products such as sulfur dioxide, particulates, or mercury that come from a coal-fired plant. However, similar to coal-fired plants, they may emit nitrogen oxides.

In 2017 there were 5,878 utility-scale natural gas turbine in the USA with a nameplate capacity of 522.4 GW. Thus the average gas-fired generator size was about 90 MW.[11] The Gila River Power Station in Arizona is a large gas-fired power plant. It has four power blocks, each with two combustion turbines and two HRSGs. Its total capacity is 2200 MW.[12]

7.2.3 Nuclear Power Plants

In some ways, nuclear power plants are similar to coal-fired power plants. Both use steam turbines connected to alternators to generate electricity. However, whereas combustion of coal is used to generate steam in coal-fired plants, heat resulting from nuclear fission is used in nuclear power plants. Since no fossil fuels are burned in a nuclear power plant, they do not emit carbon dioxide during operation.[13] Uranium is the fuel used in nuclear reactors. Uranium has two naturally occurring isotopes: ^{238}U and ^{235}U. Of these, only ^{235}U is fissile, so it is the essential fuel. Some reactors use enriched uranium with a higher proportion of ^{235}U than occurs in nature.

As nuclear fission occurs, the chemical composition of the fuel changes. After some time the fuel becomes less able to sustain a nuclear reaction, and must be disposed. The resulting spent nuclear fuel will be radioactive for hundreds of thousands of years. To put that in perspective, Jesus was alive about 2000 years ago, and the prophet Muhammad about 1400 years ago. No solution has been found for the safe storage of spent nuclear fuel due to its extremely long life.

Two types of nuclear power plants are used in the USA: boiling water reactors and pressurized water reactors. In a boiling water reactor (illustrated in Fig. 7.8), the heat from fission is used to boil water in order to produce steam that passes through a steam turbine. In a pressurized water reactor (illustrated in Fig. 7.9), heat from fission is used to heat pressurized water (shown as the coolant loop), which does not boil due to its high pressure. This hot water flows to a steam generator where it heats

---

[11] https://www.eia.gov/electricity/annual/html/epa_04_03.html

[12] https://www.srpnet.com/about/stations/gilariver.aspx

[13] A complete life cycle analysis that considers not only carbon dioxide emissions during operations, but also the carbon dioxide that is emitted during manufacture and construction of power plants, is necessary for an apples-to-apples emission comparison.

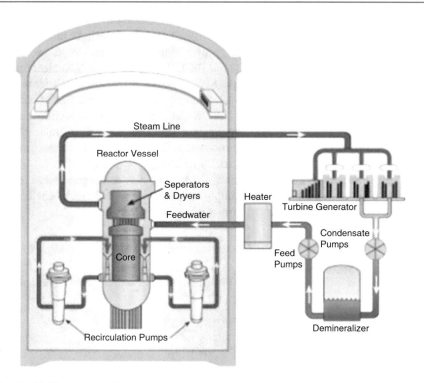

Fig. 7.8 Boiling water nuclear reactor

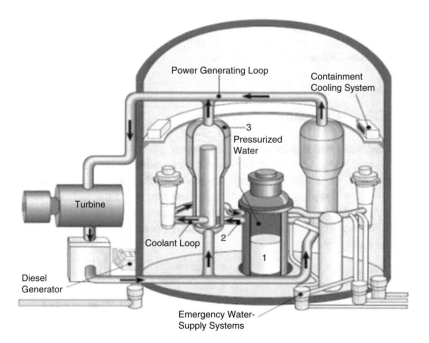

Fig. 7.9 Pressurized-water nuclear reactor

a separate stream of water (shown as the power-generating loop) at lower pressure and boils it into steam. That steam is then passed through a steam turbine.

At the end of 2019 the USA had 96 operating nuclear reactors at 58 nuclear power plants with a total capacity of about 100 GW. The Grand Gulf Nuclear Station in Mississippi has a capacity of 1400 MW. Nuclear power plants all have large capacities. The smallest plants in the USA have capacities of about 520 MW.[14]

Figure 7.3 shows that about 19% of all US electricity was generated from nuclear power. This is much more than the nuclear share of generation capacity (see Fig. 7.10). Power plants do not always operate at full capacity. For example, natural gas generated about 38% of US electricity in 2019, which is less than its 43% of generation capacity. Whereas it is relatively easy to change the generation from a gas-fired plant by adjusting the fuel supply to the plant, it is technically challenging to adjust the amount of generation from nuclear plants.

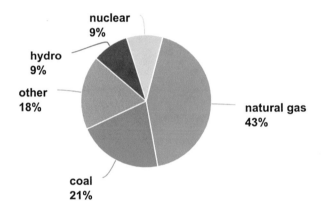

Fig. 7.10 US electrical generation capacity (2019)

7.2.4 Diesel Generators

Diesel generators are not as large as the power plants that have been described above. They are typically used in various circumstances: places without connection to a power grid, on islands, as emergency power backup in case a grid fails, or as part of a microgrid system. They consist of a diesel engine connected to an electrical generator, usually an alternator. Their power output ranges from about 8 kW to 2000 kW. Many diesel generators are mounted on trucks, and can be deployed quickly in emergencies.

---

[14] https://www.eia.gov/energyexplained/nuclear/us-nuclear-industry.php

7.2.5 Hydropower

If you have ever stood underneath a waterfall or held a high-pressure water hose, you know that water can exert a substantial force. Such a force can be harnessed by building a dam to create a lake or reservoir behind it with a higher water level upstream than downstream. Then the water can be allowed to fall through a pipe to the lower level driving a turbine to generate electricity. The energy can also be harnessed by various bucket, paddle, or turbine systems. An early use of hydropower is illustrated in Fig. 7.11. Historically, such watermills were used for milling (grinding), rolling, or hammering.

Fig. 7.11 Twelfth-century watermill

Technology has come a long way over time. Figure 7.12 shows part of the Three Gorges hydropower project in China that has a capacity of 22.5 GW, the same as that of about 23 large fossil fuel plants! But the fundamental energy concepts of Three Gorges and the watermill are the same. The potential energy in the water behind a dam is transformed into electrical energy. Recall that Fig. 7.4 shows the important role of hydropower in China's electrical generation mix.

Fig. 7.12 Three Gorges dam

7.2.6 Solar

You may have seen solar panels on the roof of a building or perhaps in a solar farm. The sun emits solar energy that makes our planet habitable. This energy can be harnessed in various ways. Photovoltaic (PV) panels on roofs or in solar farms convert the sun's energy directly to electricity. PV cells consist of semiconductors. They convert only about 5–15% of the sun's energy into electricity. The cells receive energy as solar radiation and generate direct current (DC). The electrical system discussed in Sect. 7.1.1 uses alternating current (AC) as do most appliances (except small devices such as cellphones and batteries). A device called an inverter converts DC into AC.

There are two other uses of solar energy. The thermal energy of the sun can be used to heat homes directly. Sunlight passes through windows, ideally striking inside structures such as tile floors or brick walls that warm up and then release heat later. It is useful during winter, but a nuisance during summer when a house may get too warm. Thank goodness for curtains or blinds to limit solar heating during the summer. Another thermal application of solar power is for heating water to use in homes. Solar collectors that focus the sun's rays on water are installed on the roof of a home. Hot water is then stored in a hot water tank somewhere in the house. Such thermal uses of solar energy are referred to as passive solar.

There is yet another thermal application of solar energy, known as concentrated solar power (CSP). It is used to generate electricity on a commercial scale. A CSP plant uses mirrors to focus sunlight on a receiver, where a transfer fluid is heated.

The heat in the transfer fluid is used to create steam in a separate water-steam loop, which is then used to generate electricity with a steam turbine similar to the steam cycle in coal plants or nuclear plants. As of 2019 there were four large CSP plants in the USA with capacities from 250 MW to 357 MW.[15]

Since solar energy systems do not consume fossil fuel during operation, there are no associated emissions of carbon dioxide.[16] However, solar energy systems can only operate when the sun's rays are incident on them, i.e., during the day. Also cloud cover affects the amount of solar energy that is incident on the systems. Passive solar systems are designed to store solar energy either in hot water tanks or thermal masses such as walls and tiled floors. Solar energy for CSP plants can be stored in molten salt systems. However these are not widely deployed.

The intermittent nature of solar energy generation means that it must be complemented with other types of generation, such as fossil fuel, in order to meet power demand, or with some means for storing energy (such as batteries or pumped storage). The cost of solar energy equipment has fallen dramatically over time, so its deployment is increasing rapidly.

7.2.7   Wind

Wind energy has been used for centuries (see Fig. 7.13). Early uses included grinding wheat into flour and pumping water in the Netherlands since much of the land there is at or below sea level.

Fig. 7.13 Seventeenth-century windmill

---

[15] https://www.eia.gov/energyexplained/index.php?page=solar_thermal_power_plants

[16] Energy is used in the manufacture of solar energy equipment, which may result in some emissions.

Over the past several years remarkable progress has been made at harnessing wind energy to generate electricity. The capacity of a wind turbine can range from below 100 kW for directly powering a home, farm, or small business up to several megawatts in commercial applications. Wind turbines can be located on land or offshore. The towers for large turbines stand 80 meters tall. Typically they support a rotor with three blades. The blades turn a shaft connected to a gearbox and generator. Modern turbines begin operating when the wind speed exceeds a cut-in speed, which is about 6 to 9 miles per hour. They shut down if the wind blows too hard, above about 55 miles per hour. The power generated by wind turbines varies with the cube of the wind speed, so a doubling in wind speed increases the power by a factor of 8.

There are several commercial wind farms in the USA. Due to the area necessary for wind turbines, commercial wind farms are located some distance away from load centers. The cost of electricity transmission lines that are necessary for power delivery from wind farms to load centers has hampered the development of some wind farms.

Like solar energy, no fossil fuel is used during operation of a wind turbine, so there are no emissions of carbon dioxide. However, also like solar energy, wind energy is not available all of the time. It is only available when the wind is blowing at the right speeds. Also the power output varies significantly with the wind speed. So the intermittent nature of wind energy means that, like solar, it must also be complemented with alternative types of generation or energy storage. Also like solar, the cost of wind turbines has fallen dramatically, so much so that electricity from wind is now less costly to generate than that from some other types of fuel.[17]

7.2.8 Other

Small amounts of electricity are generated with other technologies. Geothermal energy is heat from the earth. This heat can be extracted from the earth by drilling wells that produce steam. Some geothermal projects inject cold water into an injection well and produce steam from a nearby producing well. The steam is used in a steam turbine to generate electricity. There are also binary cycle geothermal plants, which transfer heat from hot water produced in a geothermal well to boil a working fluid, typically an organic compound with a low boiling point. The working fluid is used to drive a turbine, and follows a thermodynamic cycle similar to the water-steam cycle in steam generators.

Another alternative is to burn biofuels such as wood chips to generate steam, rather than using coal, natural gas, or nuclear for heating. The Drax power plant in the UK has a capacity of 2.6 GW partly powered by biomass such as wood pellets.[18]

---

[17] It is important to consider both the operating and capital costs of all energy systems for a fair comparison.

[18] https://en.wikipedia.org/wiki/Drax_Power_Station

There is a 50 MW generation unit at Schiller Station in Portsmouth, New Hampshire, that uses wood chips as fuel.[19]

7.3   Transmission and Distribution

Most power plants are distant from demand centers so transmission lines are used to carry their electricity. There are about 360,000 miles of electricity transmission lines in the USA.[20] Energy losses occur in transmission lines. Such losses can be reduced by carrying the electricity at a high voltage. It is easy to change the voltage of alternating current by use of a transformer. Electricity is generally produced in power plants at 5 to 34.5 kilovolts (kV).[21] Voltage is increased at the outlet of power plants in step-up transformers. Transmission lines operate at various voltages from 138 kV to 765 kV.

Transmission lines have the capacity to deliver large amounts of power but are expensive to build. Within cities and neighborhoods it is economical to distribute power using lower voltage distribution networks rather than transmission lines. Distribution networks operate at various voltages, often between 7 and 72 kV. Step-down transformers reduce the voltage from transmission lines for distribution networks. Residential customers in the USA receive power at 120 V or 240 V. Additional transformers are used in the distribution network to reduce the voltage to 120 V and 240 V. Many industrial customers have high power demands. In this case they receive high-voltage power directly from either a transmission line or a distribution line.

7.4   Efficiency

7.4.1   Generation

Fossil fuel power plants convert chemical energy into electrical energy. The standard measure of conversion between the two is known as the heat rate, expressed in BTU/kWh. This is the amount of energy in BTUs that is required to produce 1 kWh of electrical energy. Lower heat rates mean that less input energy is required to produce the same amount of electrical energy. The average heat rate for US power plants in 2017 is shown in Table 7.1. The heat rates for steam generators are about the same regardless of the type of fuel. This is because the steam cycle in such plants is the same for each type of fuel. Combined cycle plants have the lowest heat rates. The use of both combustion turbines and steam turbines in such plants results in better use of the chemical energy in the fuel.

---

[19] https://www.power-technology.com/projects/wood-schiller/

[20] Ibid page 13.

[21] https://www.energy.gov/sites/prod/files/2015/12/f28/united-states-electricity-industry-primer.pdf page 12.

Table 7.1 US power plant
heat rates

BTU/kWh	Coal	Petroleum	Gas	Nuclear
Steam generator	10,043	10,199	10,353	10,459
Gas turbine		13,491	11,176	
Internal combustion		10,301	9,120	
Combined cycle		9,811	7,649	

Heat rates can be used to determine the efficiency of the energy conversion from fuel to electricity. Both BTUs and kWh are measures of energy. 1 kWh is equivalent to 3412 BTU. So the efficiency of a power plant is 3412/heat rate (in BTU per kWh). The efficiencies corresponding to the heat rates of the previous table are shown in Table 7.2.

Table 7.2 US power plant
efficiencies

Efficiency	Coal	Petroleum	Gas	Nuclear
Steam generator	34%	33%	33%	33%
Gas turbine		25%	31%	
Internal combustion		33%	37%	
Combined cycle		35%	45%	

Note that only a quarter to less than half of the energy provided in the fuel is delivered in the form of electricity. It may seem surprising that the efficiencies are so low. Steam generators have an efficiency of only about 1/3. That is, the electrical energy output is only 1/3 of the chemical energy that is consumed as input. Recall from Chap. 2 that energy is conserved. The bulk of the energy that is supplied to a power plant is wasted in the form of heat. You may wonder why such low efficiencies are tolerated. It is beyond the scope of this book to explain, but the laws of thermodynamics are a primary determinant of the efficiencies. Even though thermodynamics limits the efficiency, you may still wonder why such energy losses are accepted. Why bother to waste most of the energy in the fuel to generate electricity? To answer this question, one must consider the alternative. If energy is not delivered in the form of electricity then it must be delivered in some other form. Electricity is a truly convenient form of energy. A user simply flicks a switch, although the process is a bit more involved for industrial customers! Efficiency losses are tolerated because electrical energy is so easy to use.[22]

The heat rates and efficiencies shown in the two tables are *averages* for the USA. This means that there are some plants with heat rates below the average and others above it. In particular, there are some large combined cycle power plants that have efficiencies of almost 60%, even though the average efficiency is only 45%.

---

[22] There is another consideration. The efficiency of many devices such as engines that use energy in other forms besides electricity is also low, whereas the efficiency of conversion of electrical energy into energy services can be quite high. Thus efficiency losses are inevitable. They occur either in the end device or during generation of electricity.

7.4.2   Transmission and Distribution

Generation is not the only place where energy losses occur. Energy losses also occur in the transmission and distribution networks. Losses in transmission and distribution lines depend on the length of the lines and also on the electrical current in the lines. The step-up electrical transformers shown in Fig. 7.1 have higher output voltages and lower output currents than the inputs. The lower currents reduce losses in the lines. But high voltages are not suitable for most end users, so step-down transformers are used near load centers to reduce the voltage. When the voltage is reduced the current and energy losses increase. About 5% of US electricity generation is lost in transmission and distribution.[23]

7.5   Emissions

Most of the different generation technologies discussed above cause emissions of various products such as carbon dioxide, water, sulfur dioxide, nitrous oxides, mercury, and particulates. Renewable energy (solar and wind) and hydropower are exceptions. They do not have such emissions during operation. All of the emissions, except water, have adverse environmental consequences. Carbon dioxide has been linked to climate change, sulfur dioxide causes acid rain, nitrous oxides create a brown cloud that limits visibility and is injurious to human health, mercury can cause brain damage, and particulates can cause respiratory illnesses. You might wonder why such adverse impacts are tolerated. This is a really important question for public policy. Electrical energy provides huge social benefits. Are negative health and environmental impacts acceptable as a consequence of the energy benefit? It is up to you to debate this and decide, or to figure out better alternatives such as increased use of renewables and/or more efficient use of energy. The answer is likely a combination of several things, including more use of renewables. But the intermittent nature of renewable energy requires a technical solution such as energy storage or a backup supply, perhaps fossil fuel generation.

To assist in your thinking and debate, recall that Sect. 2.5.5 discussed combustion of fuels and the resulting emissions of carbon dioxide. Table 7.3 repeats some data about those emissions from Table 2.10 and includes representative heat rates from Table 7.1 to determine emissions of carbon dioxide per kWh of electricity generation.

The table shows that the lowest rate of emissions per kWh is from a combined cycle natural gas plant. Such a plant has two advantages: (1) the carbon content in the natural gas fuel is relatively low, so CO_2 emissions per MMBTU are low, and (2) the efficiency of energy conversion is relatively high (as evidenced by the low heat rate). The highest rate of emissions is from a coal plant. The heat rate and efficiency of such a plant are not much different than those for a diesel generator or simple cycle natural gas plant, but the CO_2 emissions per MMBTU are high due to the high

---

[23] https://www.eia.gov/tools/faqs/faq.php?id=105&t=3, Retrieved Feb. 21, 2019.

Table 7.3 Carbon dioxide emissions for various fuels

Fuel	CO_2 emissions, lb. per MMBTU	Technology	Heat rate, BTU/kWh	CO2 emissions, lb. per kWh
Natural gas	117	Internal combustion[a]	10,301	1.21
		Simple cycle	11,176	1.31
		Combined cycle	7,649	0.89
Diesel	161	Internal combustion	10,301	1.66
Coal (anthracite)	210	Coal	10,043	2.11

[a]It is possible to use natural gas as fuel in an internal combustion engine

carbon content in coal. The carbon dioxide emissions per kWh from a coal plant are more than twice as high as those from a combined cycle natural gas plant. Figure 6.10 shows that coal-fired power plants are being taken out of service in many countries. Table 7.3 shows that such retirements can reduce emissions of CO_2 even if the generation capacity of those coal plants is replaced with plants using other technologies.

7.6   Electricity Consumption

7.6.1   Consumption by Sector

Total US electricity generation was 4,170 TWh in 2018.[24] Total US electricity sales were 3,861 TWh,[25] which corresponds to an average power sales rate of about 441 GW for the entire country. However, the average rate is not particularly meaningful, since the power consumption varies significantly by season and by time of day, as discussed below in Sect. 7.6.2.

Figure 7.14 shows that the residential sector consumed the largest amount of electricity in the USA during 2018. The commercial sector was a close second, with the industrial sector being third, but still a significant consumer with 25% of electricity sales. Transportation does not appear in the figure since penetration of electric vehicles is small.

Figure 7.15 provides some detail about residential uses of electricity. It shows that household appliances (shown as "all other uses") such as clothes washers and driers use substantial amounts of electricity. Refrigerators, freezers, and televisions also use significant amounts of energy, but not as much as space cooling, i.e., air-conditioning. Water heating and space heating use lesser amounts: 9.5% and 8.5% of residential electricity consumption, respectively. This low usage may seem to

---

[24] https://www.eia.gov/tools/faqs/faq.php?id=74&t=11#:~:text=In%202018%2C%20total%20 U.S.%20electricity,of%20CO2%20emissions%20per%20kWh

[25] Calculated from 2019 and prior year change numbers at https://www.eia.gov/energyexplained/ electricity/electricity-in-the-us-generation-capacity-and-sales.php

Fig. 7.14 US electricity
sales by sector (2018)

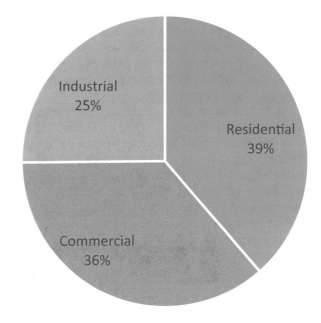

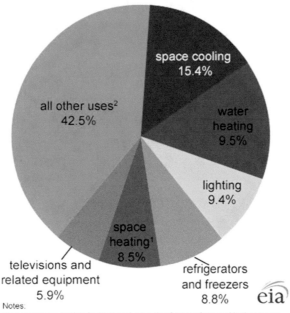

Fig. 7.15 US residential electricity consumption

contradict the fact that water heating and space heating were about 40% of residential energy use (see Fig. 3.7). The seeming contradiction is resolved by recognizing that both natural gas and electricity are used for water heating and space heating in the residential sector. Natural gas provides a significant amount of energy for both. Since natural gas is used for those purposes in many residences, the share of electricity used in residences is lower.

Figure 7.16 shows the corresponding breakdown for electricity use in the commercial sector. As you would expect the commercial sector consumes a large share of its electricity for computers and office equipment. It also uses a significant amount of electricity for ventilation, which does not require electricity in most residential settings. Windows cannot be opened in many office buildings so ventilation can only be achieved by blowing air, which requires electricity for motors. The share of electricity used in the commercial sector for space and water heating is much lower than in the residential sector. It is rare to use electricity for space heating in commercial buildings. Central heating systems that use natural gas or fuel oil are more common.

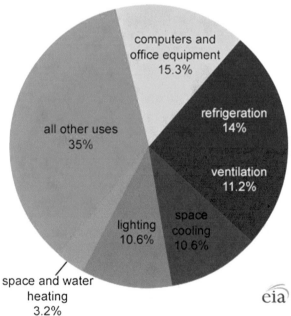

Source: U.S. Energy Information Administration, *Annual Energy Outlook 2018*,
Table 5, February 2018

Fig. 7.16 US commercial energy consumption

Figure 7.17 shows the corresponding data for shares of electricity consumption in US manufacturing. As you would expect, the shares are very different than in the residential and commercial sectors. The largest share is consumed by machine drives, i.e., motors. Process cooling, refrigeration, and heating also consume large shares. Many types of manufacturing use cooling and heating in their processes. Some manufacturing types use electrochemical processes. These processes are used to make chlorine and caustic soda, as well as in the production of aluminum.[26]

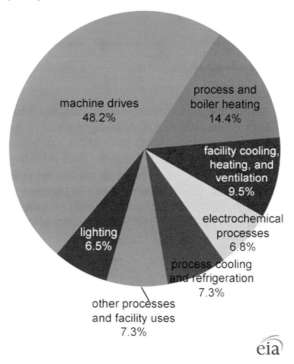

Fig. 7.17 US manufacturing electricity consumption

7.6.2 Fluctuations

Demand for electricity fluctuates by season and also by time of day. Figure 7.18 shows monthly power demand for three sectors in the USA during 2018. Power demand in each of these sectors fluctuates by month; however the fluctuation in the industrial sector is the smallest reflecting the almost steady operation of industrial processes over the course of a year. By contrast the demand for power in the

---

[26] https://www.electrochem.org/dl/interface/fal/fal14/fal14_p049_055.pdf

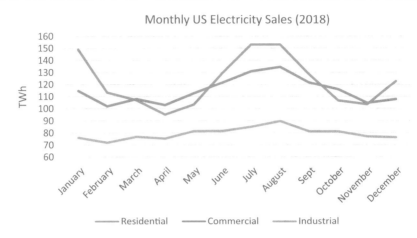

Fig. 7.18 Monthly US electricity sales

residential and commercial sectors fluctuate significantly, and is highest in the summer. In the residential sector it is also high during the winter. Residential power demand is high during the summer due to increased use of air-conditioning. It is high during the winter due to increased demand for electric heating and lighting.

The seasonal variation in power demand is not the same in all US states. Figure 7.19 compares the monthly power demand in Montana and Nevada. The summer demand is high in Nevada, a desert state with extensive air-conditioning. By contrast the winter demand is high in Montana, a northern state, due to the large demand for electric heating during cold winters.

Fig. 7.19 Different monthly demands (2018)

Power demand fluctuates not only seasonally, but also minute to minute. Figure 7.20 shows the hourly power demand during 5 days at the PJM regional transmission organization which coordinates movement of wholesale electricity in 13 northeastern states in the USA and the District of Columbia. Several features are notable. Power demand was lower on March 16 and 17 compared to later days in the figure. These were weekend days. Also, there are two peaks in demand during the course of each day: one in the late morning, and the other in the late evening. Correspondingly there is low demand in the early morning, during the midafternoon, and at night. This fluctuating demand is a reflection of how residential customers use electricity. They use electricity in the morning as they get ready for their days, and in the late afternoon/early evening for cooking and other activities. The electrical system must accommodate both time-of-day and seasonal fluctuations in demand.

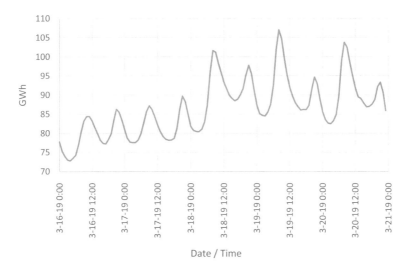

Fig. 7.20 Hourly power demand—PJM

7.6.3   Residential Power

Table 7.4 provides details about the power consumption of various residential appliances. Recall that power is the rate at which energy is consumed (or supplied). The table also shows estimates of both the energy consumed per year by each appliance based on an estimate of how many days it is used per year, and for how many hours during each use, and the instantaneous power use by the appliance. For example, the instantaneous power drawn by a home air conditioner is between 1000 W and 4000 W. If the unit draws 2500 W on average and runs for 12 hours per day on 90 summer days per year, then the energy it consumes is 2500 W × 12 hours per day × 90 days = 2700 kWh in a year. As another example, a 100 W incandescent light bulb draws 100 W. If it is used for an average of 4 hours per day for 365 days, then it uses 100 W × 4 hours per day × 365 days = 146 kWh per year. A clock radio uses very little power, between 1 and 2 W. If it uses 1.5 W for 24 hours per day for 365 days per year, then it uses 13 kWh during a year.

Table 7.4 Energy consumption of household appliances

	Power consumption[a] W			Hours per	Days per	Annual energy,
	Min	Max	Standby	day	year	kWh[b]
Home air conditioner	1000	4000	N/A	12	90	2700
Fridge/freezer	150	400	N/A	24	365	2409
Refrigerator	100	200	N/A	24	365	1314
Electric heater fan	2000	3000	N/A	4	90	900
Gaming PC	300	600	1	4	365	657
Clothes dryer	1000	4000	N/A	4	60	600
Desktop computer	100	450	N/A	8	260	572
Coffee maker	800	1400	N/A	1	365	402
Freezer	30	50	N/A	24	365	350
Oven	2150	2150	N/A	1	150	323
Dishwasher	1200	1500	N/A	2	104	281
Electric stove	800	1000	N/A	1	300	270
Laptop computer	50	100	N/A	8	365	219
Hair blow-dryer	1800	2500	N/A	0.5	180	194
Game console	120	200	N/A	3	365	175
100 W Light bulb (incandescent)	100	100	0	4	365	146
Electric blanket	200	200	N/A	8	90	144
Submersible water pump	360	400	N/A	1	365	139
Microwave	600	1700	3	0.3	365	126
Home sound system	95	95	1	3	365	104
Espresso coffee machine	1300	1500	N/A	0.2	365	102
60 W Light bulb (incandescent)	60	60	0	4	365	88
Home Internet router	5	15	N/A	24	365	88
Electric kettle	1200	3000	0	0.1	365	77
Ceiling fan	25	75	0	6	180	54
Iron	1000	1000	N/A	1	52	52
Washing machine	500	500	1	1	100	50
Phone charger	4	7	N/A	24	365	48
Toaster	800	1800	0	0.2	180	47
Vacuum cleaner	200	700	0	1	52	23
Clock radio	1	2	N/A	24	365	13
LED light bulb	7	10	0	4	365	12
Guitar amplifier	20	30	N/A	2	180	9
Heated bathroom mirror	50	100	N/A	0.3	365	8
Inkjet printer	20	30	N/A	1	300	8
Table fan	10	25	N/A	2	120	4
Straightening Iron	75	300	N/A	0.2	100	4

[a]https://www.daftlogic.com/information-appliance-power-consumption.htm Retrieved Feb. 21, 2019
[b]Assuming power consumption at average of minimum and maximum

There is wide variation in the energy consumption of different appliances. Air conditioners, fridges, freezers, and electric heater fans use large amounts of energy. Inkjet printers, LED lightbulbs, and clock radios use hardly any energy. Toasters have high instantaneous power consumption, but do not use much energy over the course of a year if they are not used too much.

Note that the standby power consumption of appliances such as microwaves is very low. So you do not need to worry about wasting too much energy if your appliance is in standby mode.

You may wonder about the power consumption of televisions. The table below shows the operating power consumption for about 3600 types of TV with different sizes and technologies. It shows that (1) larger TVs use more power, (2) LCD TVs use the least power, and (3) plasma TVs use the most power.

	Average power, W			
Size	LCD	LCD (LED)	OLED	Plasma
L	110	122	149	163
M	46	48	89	108
S	25	21		

Size (diagonal inches): S < 30, M: 30 to 48, L > 48
LCD (LED) is LED backlighting with liquid crystal display
OLED is organic LED

7.7   Electricity Storage

Section 7.6.2 explains how electricity consumption varies seasonally and by time of day. Electrical systems should supply the amount of power that is needed at every moment. If there is insufficient capacity, then power will need to be rationed. Some users will be unable to receive the power that they demand. There are various possible solutions for this. In some countries, certain power customers can sign up for interruptible service. If there is a shortage of power in the system, then power for interruptible customers is cut off. Customers pay a reduced rate for such interruptible power. Another possible solution is to build a power system so that it is guaranteed to have sufficient capacity. Since it is difficult to predict power demand accurately one would typically include excess capacity above the predicted maximum demand.

It is costly to build power capacity. So building capacity to meet or exceed predicted peak demand is expensive. Furthermore it is wasteful, since the peak demand occurs only for a brief period of time. Thus most of the time there would be unused capacity. One way to avoid underutilized capacity is to interconnect with other electrical systems, in the hopes that one system can draw power from another when it is short, and vice versa. Such interconnections do exist. Careful technical and economic coordination is required for their success.

Another seemingly obvious solution is to develop storage for electrical energy. With storage, energy can be stored when end-user demand is less than the system capacity, and withdrawn when demand exceeds capacity. Such a storage scheme is used for natural gas, since it is relatively easy to store gas by injecting it into underground reservoirs and to withdraw it when needed. However there are technological limitations to the ability to store electrical energy. Section 2.4 describes methods that exist for storing energy that can be withdrawn as electricity. The capacity of such systems is limited by geography in the case of pumped storage and by fundamental technical issues for batteries. See Question 12 for an illustration of how the size of available battery storage is very small compared to generation capacity.

The ability to store large amounts of electrical energy would be so useful that it is a very active research area. Battery technology has come a long way. Improvements have been crucial for the widespread adoption of cellphones. But the power used in a cellphone is tiny compared to that required for significant impacts in the electrical grid. The increasing use of batteries to power cars is also evidence of improvements in battery technology. (See Table 2.5 for a comparison of battery and liquid fuel energy storage in cars.) Many possible new battery technologies are being explored, as well as hybrid-type storage technologies. For example, some people are considering the use of excess electricity to extract hydrogen from water during off peak times and then to use the hydrogen during peak periods to generate electricity. If renewable energy is used to extract hydrogen, it is referred to as "green hydrogen." This is in contrast to the situation in which fossil fuels are used. Burning of the fossil fuels causes emissions of carbon dioxide, a negative environmental consequence.

Storage will also have a large impact on the future use of renewable energy. The power output of solar cells depends on the solar incidence. So clouds and nighttime limit their output. Also, the power output of wind turbines is very sensitive to the wind speed, which is always variable. A large-scale energy storage technology will be very useful to accommodate the intermittent nature of renewable energy generation.

The cost of electricity storage has plummeted in recent years. This has led to the use of huge utility-scale batteries. A battery system with capacities of 300 MW and 1200 MWh is being constructed at Moss Beach, California. This system is an alternative to new gas-fired plants for meeting peak power demand that would otherwise be required.

Batteries provide other grid services besides storage. They respond almost instantaneously to changes in power demand, unlike other generation technologies such as coal-fired or nuclear power plants. This flexibility is useful for maintaining the frequency of the alternating current that is used in grids, and also for black start capability that is required when a power plant has gone down and needs to be restarted.

7.8   Kettle Example

Many energy concepts can be illustrated by considering how an electric kettle boils water. Just in case you are not familiar with an electric kettle, water is added to the kettle which is plugged into an electric socket. After switching on the kettle, power from the wall goes to the kettle, warming the water. As the water heats, the walls of the kettle become hot, and heat is lost to the air around the kettle. This is illustrated in Fig. 7.21. The figure shows bubbles in the kettle. These are bubbles of water vapor, i.e., steam, in the liquid water. The water that was added to the kettle in liquid form becomes steam (the vapor form) when the temperature of the water reaches the boiling point. Read on, and you will understand why allowing the water to continue boiling is a waste of energy!

Fig. 7.21 Kettle energy flows

Each kettle consumes power at a roughly constant rate determined by the design of its heater. Recall that energy is power multiplied by time. The heat that is added to the kettle is this amount of energy. The longer the kettle is on, the more energy it consumes. Heat is lost from the kettle when the water is hotter than the surrounding air. The heat loss is proportional to the difference between the water temperature and the temperature of the outside air. When the heat that is added to the kettle is greater than the heat loss, the water inside the kettle heats up. Thermal energy is stored in the heated water. The energy stored in the water is equal to the mass of water multiplied by the specific heat of the water, a physical property of the water, multiplied by the temperature of the water.

Since energy is conserved, the rate of change in the thermal energy equals the difference between the rate at which energy is supplied to the kettle and the rate at which energy is lost from the kettle. The appendix to this chapter solves the differential equation that characterizes the energy flows and temperature change. You do not need to read the appendix to understand what follows.

This kettle discussion is valid when there is only liquid water in the kettle. However, when the temperature of the water reaches its boiling point then there are still energy flows into and out of the kettle, but a new thing happens. The water undergoes a phase transition. It changes from liquid water to water vapor, known as steam. Energy is required to convert any liquid to vapor. This energy is known as the latent heat of vaporization. The temperature of the water remains constant while liquid water is vaporizing into steam. Since you can only drink liquid water (or your favorite beverage using that liquid water) not steam, there is no gastronomic benefit associated with producing steam.[27] Extra energy input does not change the water temperature when it is boiling; it simply converts liquid water into steam. The latent heat of water is 2260 J/g, so each kilojoule that is added to the contents of a boiling kettle creates 1000 J/2260 J/g = 0.44 g of steam. Since kettles are not sealed, steam escapes from a kettle when it is boiling. It would boil dry if left operating for too long.[28]

Before reading this section, you probably never spared a thought about boiling a kettle. Now you know that the energy transferred from the electrical socket heats the water, resulting in thermal energy stored in the water. Some of the heat that is applied to the water is lost to the surrounding air. Furthermore, the hotter the water gets, the quicker the rate at which heat is lost to the surroundings. Can you see this in the red line in Fig. 7.22, which shows how the water temperature changes over time?

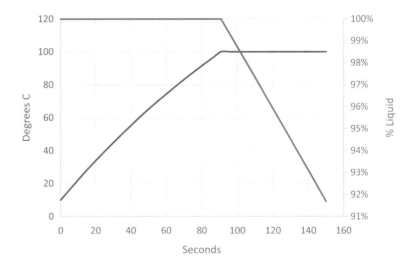

Fig. 7.22 Model of electric kettle behavior

---

[27] I live in Colorado, where the air is extremely dry. Residents of Colorado often add moisture to the air to compensate for this, so there is some benefit to creating the steam here!

[28] Many kettles have automatic cutoff systems to avoid boiling dry.

The temperature rises for about the first 90 s. The slope of the red curve, i.e., the rate of change of the temperature, gets lower over time. This is because the hotter the water gets, the higher the rate at which the kettle is losing heat to the air. Since the net rate at which energy is transferred to the water is the difference between the energy supplied from the socket and the rate of heat loss, higher rates of heat loss mean lower rates of net energy transfer to the water. Consequently the rate of change in temperature falls over time, even though the temperature itself is rising.

The figure shows that something dramatic happens at about 90 s. The temperature stops increasing (the red line becomes horizontal) and the percentage of liquid water in the kettle drops from 100% to something less, as the liquid is converted to steam, i.e., vapor. A simpler way of saying this is that the water is boiling! At about 150 s, i.e., 60 s after boiling begins, around 8% of the liquid water has been converted to steam. The water that remains is still at the same temperature, but energy has been wasted in the creation of steam. Moral: It is better to turn your kettle off as soon as it boils; this will not only save energy, but also lower emissions of carbon dioxide at a power plant somewhere if your electricity is generated by a fossil fuel such as coal or natural gas.

7.9   Summary

Electricity is a very convenient form of energy. It is created by converting some other form of energy into electricity. The chemical energy in fossil fuels (natural gas, coal, diesel, or wood) is a common energy source. Nuclear energy, solar, wind, and potential energy from water stored behind dams are other sources. The conversion process is not very efficient (except for hydro); only a quarter to a half of the energy supplied is converted to electricity. There are thermodynamic reasons for this.

In most cases the chemical energy in fossil fuels or nuclear energy is used to create high-pressure steam that drives a turbine connected to an alternator. The alternator generates alternating current (AC) electricity. By contrast batteries and solar panels provide electricity as direct current (DC). Power that is generated in power plants is sent to demand centers over high-voltage transmission lines. There is an inverse relationship between voltage and current, so high voltage corresponds to low current. Energy losses are lower when current is lower. Transformers are able to step up voltage at power plants, and to step it down at demand centers. Distribution lines are used to distribute electricity within towns. They operate at intermediate voltages, lower than that on transmission lines and higher than the voltage delivered to homes.

Both the sun and wind have been harnessed to generate electricity. Wind energy has been used for centuries, but the technology has improved considerably. It is now cheaper to generate electricity from wind and solar than from other fuels. However, such renewable energy has a major drawback. It is intermittent, since it relies on sunshine and wind, both of which vary over the course of a day and night. There has been rapid progress in the technology for storing electrical energy in batteries. But more remains to be done. In the meanwhile a complementary energy source such as fossil fuel is required to ensure that electricity supply is sufficient to satisfy demand.

Electricity demand fluctuates both by season and by time of day. More electricity is required in the summer for air-conditioning, and more in the winter for heating and lighting. These seasonal effects are different in locations with different climates, desert compared to mountainous. Residential energy demand tends to be high in the morning and evening when people are active at home, whereas it is low in the midafternoon and in the middle of the night. Industrial energy demand tends to be more constant over different seasons and during the course of a day since industrial processes typically operate at a steady state.

7.10 Questions

1. Figure 7.3 shows that 38.6% of US electricity was generated using natural gas in 2019. Figure 7.10 shows that 43% of the US generation capacity was natural gas. Total US generation capacity was about 1110 GW in 2019, and total generation was about 4100 TWh.

   (a) There are 8760 hours in a year. If gas-fired power plants operated at a constant rate in every hour, what fraction of their capacity was used?
   (b) Gas-fired power plants do not operate at a constant rate in every hour. As an arithmetic exercise, suppose that they either operate at full capacity or do not operate at all. What fraction of the time would they operate at full capacity?
   (c) Gas-fired power plants do not operate only at full capacity or not at all. Sometimes they operate at partial capacity. Describe a gas-fired power plant operating scenario that is consistent with the data.

2. Figure 7.3 shows that 19.4% of US electricity was generated using nuclear in 2019. Figure 7.10 shows that 9% of the US generation capacity was nuclear. Total US generation capacity was about 1110 GW in 2019, and total generation was about 4100 TWh.

   (a) There are 8760 hours in a year. If nuclear power plants operated at a constant rate in every hour, what fraction of their capacity was used?

3. Verify the efficiency numbers shown in Table 7.2.
4. Verify the CO_2 emission numbers shown in Table 7.3.
5. Section 7.6.1 states that 4,170 TWh of electricity was generated in the USA in 2018, while 3,861 TWh was sold. What fraction of the amount generated was sold? What happened to the remainder that was not sold?
6. Use the data in Fig. 7.3 and Table 7.3 to estimate the amount of CO2 emissions (in metric tons) from power plants in the USA in 2019. Total power generation was 4,401 TWh[29] in that year. Neglect emissions from the "other" and "oil"

---

[29] Source: [1].

categories. Assume that generation from natural gas was a 50/50 mix of simple cycle and combined cycle (no internal combustion).

7. Verify the statement in Sect. 7.6.1 that an average power sales rate of about 441 GW corresponds to annual sales of 3861 TWh.

8. Figure 3.7 in the text shows a breakdown of US residential energy consumption by type of use. It includes both gas and energy supplied to residences. Figure 7.15 shows a similar breakdown just for US residential electricity consumption. The data from both figures can be well matched if electricity was 72.3% of the total residential energy supply. The table below is based on these data. Column 3 shows that TV and related equipment used 4.2% of total residential energy. All this energy is supplied by electricity and none by gas. Space heating used 27.3% of residential energy. Columns 4 and 5 show that electricity and gas supplied 6.0% and 21.3%, respectively, adding up to 27.3%.

				Share of energy supplied by use			
Column #		1	2	3	4	5	6
Use		Only elec?	% of electricity use (Fig. 7.15)	Total (Fig. 3.7)	Elec	Gas	Gas/Total
TV and related equip		Y	5.9%	4.2%	4.2%	0.0%	0.0%
Space heating		N	8.5%	27.3%	6.0%	21.3%	78.0%
Fridges and freezers		Y	8.8%	6.3%	6.3%	0.0%	0.0%
Lighting		Y	9.4%	7.2%	7.2%	0.0%	0.0%
Water heating		N	9.5%	13.1%	6.7%		
Space cooling		Y	15.4%	11.8%	11.8%	0.0%	0.0%
Other		Y	42.5%	30.1%	30.1%	0.0%	0.0%
Total			100.0%	100.0%	72.3%	27.7%	27.7%

(a) Use column 4 to calculate the share of all residential electricity that was used by TV and related equipment. Hint: Your answer should be close to the 5.9% shown in column 3.

(b) Determine the value that should be entered in the blank cell in column 5 for the share of energy supplied by gas for water heating.

(c) Determine the value that should be entered in the blank cell in column 6 for the share of energy supplied by gas for water heating relative to the total energy for that use.

(d) Comment on the share of gas that is used for space heating and water heating relative to total energy use for each of those purposes.

(e) What is the relative use of gas to electricity from the totals for columns 5 and 4? Table 1.1 shows that my house in the Colorado Mountains used more than twice as much energy from gas as from electricity. Why do you think my relative energy use is so different than for the USA as a whole? Hint: My house does not have air-conditioning.

9. The power demand shown in Fig. 7.19 is much higher than that shown in Fig. 7.20 even though each chart in the former is for just 1 state, while that in the latter is for 13 states and the District of Columbia. The monthly demand for each state in Fig. 7.19 is about 10,000 GWh. Calculate the average hourly

demand, assuming 30 days in a month. How does this compare with the hourly demand shown in Fig. 7.20?

10. This question relates to fluctuating electricity demand.

    (a) What is the lowest hourly electricity consumption shown in Fig. 7.20?
    (b) What is the highest?
    (c) What capacity of the electricity system would be required to satisfy demand in every hour?
    (d) Comment on the utilization of the electrical system over time.

11. Verify the annual energy consumption shown in Table 7.4 for a fridge/freezer, an electric blanket, and a home Internet router. Assume that the power consumption is the average of the minimum and maximum values shown in the table.

12. This question is a stylized illustration of the benefit of electricity storage in batteries. Suppose that there are two demand periods in an electrical system: pre-peak and peak. The pre-peak period precedes the peak period. Suppose that the pre-peak lasts for 10 hours and the peak for 4 hours, and that the demand is 30 GW and 50 GW during the two periods, respectively. Assume that demand is constant during each period.

    (a) If there is no electricity storage, how much generation capacity is required to satisfy demand at all times?
    (b) If there is 5 GWh of electricity storage, which can be charged and discharged at any rate, how much generation capacity is required to satisfy demand at all times?

13. The Tesla battery system installed in Australia that is described in Sect. 2.4 can store 129 MWh and discharge at up to 100 MW. Assume (unrealistically) that there are no efficiency losses.

    (a) Figure 7.20 shows that PJM power demand in many hours was 90 GWh. Suppose that a power plant on that system with a capacity of 1 GW went down during such an hour. What would be the shortage in energy during that hour?
    (b) What fraction of the lost capacity could the Tesla battery system supply if it was connected to the PJM grid?
    (c) How long could the battery supply power based on its capacity if it discharged at 100 MW?
    (d) Would such a battery system provide adequate backup?

14. This question is about rationing. The second part may resonate with you if you have ever experienced a power blackout.

    (a) Suppose that you live with three other people, so a total of four people including you.

        (i) Either make up a brief profile of the four people (e.g., age, gender, health) or describe an actual situation.

(ii) Suppose that food has been scarce, and that you have come across a one-ounce piece of delicious dried fruit. Describe how you would share the fruit among the four people and why.

(b) Suppose that you are responsible for providing electrical power to a small developing economy that has residential and industrial sectors. Suppose also that there is insufficient electricity generation capacity to satisfy all of the power demand for each of the sectors all of the time. Describe how you would allocate the available power in your economy. Consider every day of the week and every hour of the day.

15. (Mathematical) Solve the differential equation in the appendix.

Kettle Appendix (Mathematical)

The behavior of the electric kettle described in Sect. 7.8 can be modeled by a differential equation. The thermal energy of the water in the kettle is $mc_p T$, where m is the mass of water, c_p is the specific heat of the water, and T is the temperature of the water. Let Q be the constant rate at which electrical energy is supplied to the kettle, which is the same as the power supply of the kettle. The rate at which heat is lost from the kettle is proportional to the difference between the temperature of the water and the temperature of the air that surrounds it, T_A. Let k be the constant that characterizes the heat loss, so the rate of heat loss is $k(T - T_A)$. The net rate of energy added to the water in the kettle is the rate at which energy is added less the rate of heat loss, i.e., $Q - k(T - T_A)$.

The rate at which the thermal energy increases is equal to the net rate of energy addition to the kettle, so

$$\frac{d\,mc_p T}{dt} = Q - k\left(T - T_A\right).$$

This equation is valid until the water reaches its boiling point. The mass of water in the kettle remains almost constant until the water boils.[30] The specific heat changes slightly with temperature, but it is a reasonable approximation to assume that it is constant. Thus the differential equation can be rewritten as

$$mc_p \frac{dT}{dt} = Q - k\left(T - T_A\right).$$

The solution of this equation is

$$T = T_0 e^{-\alpha t} + \left(\frac{Q}{k} + T_A\right)\left(1 - e^{-\alpha t}\right),$$

---

[30] A negligible amount of water is lost as vapor until the water boils.

where $\alpha = \dfrac{k}{mc_p}$, and T_0 is the initial temperature of the water. The left portion of Fig. 7.22 in the main body of the text (during which the temperature increases and there is 100% liquid) shows a solution of this equation. Figure 7.23 illustrates the energy balance for the kettle. The blue line shows the cumulative amount of energy applied to the kettle over time; the orange line shows the thermal energy contained in the water over time (with a base line of zero for the thermal energy at the initial temperature of the water). Over about the first 30 s the thermal energy rises slightly more than the amount of electrical energy supplied. During this time the air temperature is higher than the water temperature by assumption,[31] so some heat is transferred from the room to the water. After about 30 s, the water temperature is higher than the air temperature, so heat is lost to the air. By the time the water boils, about 17% of the electrical energy that has been supplied to the kettle has been lost to the air, as evidenced by the orange line being below the blue line at the end of the curves.

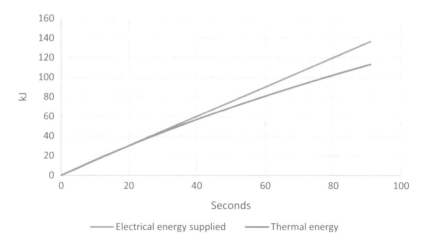

Fig. 7.23 Kettle energy

Sources

General: [1] BP Statistical Review of World Energy, June 2020

Figures

Number	Source
7.1	https://www.ucsusa.org/clean-energy/how-electricity-grid-works#.XG7HiJNKgWo
7.2	https://www.solarpowerauthority.com/how-much-does-it-cost-to-install-solar-on-an-average-us-house/

---

[31] The solution shown assumes that $\alpha = 0.0557$, $k = 7$ W/°C, $Q = 1500$ W, $T_0 = 10$ °C, $T_A = 22$ °C.

Number	Source
7.5	Adapted from https://articles.marketrealist.com/2015/01/industry-structure-analysis-coal-fired-power-generation-equipment/
7.6	http://eps-e.ru/en/b-gradirni/
7.7	https://www.entrepose.com/en/vinci-environnement/our-expertise/gas-fired-thermal-power-plants/
7.8	https://www.eia.gov/energyexplained/index.php?page=nuclear_power_plants#tab2
7.9	https://www.eia.gov/energyexplained/index.php?page=nuclear_power_plants#tab2
7.10	https://www.eia.gov/energyexplained/nuclear/us-nuclear-industry.php
7.11	By Jean-Pol Grandmont—travail personnel (own work), CC BY 2.0, https://commons.wikimedia.org/w/index.php?curid=191210
7.12	https://www.chinahighlights.com/yichang/attraction/three-gorges-dam.htm
7.13	https://commons.wikimedia.org/wiki/File:Brill_windmill_April_2017.jpg
7.14	US EIA
7.15	https://www.eia.gov/energyexplained/index.php?page=electricity_use
7.16	https://www.eia.gov/energyexplained/index.php?page=electricity_use
7.17	https://www.eia.gov/energyexplained/index.php?page=electricity_use
7.18	US EIA
7.19	US EIA
7.20	US EIA

Table

Number	Source
7.1	https://www.eia.gov/electricity/annual/html/epa_08_02.html
7.4	https://www.daftlogic.com/information-appliance-power-consumption.htm

Petrochemicals

8

8.1 Plastics and More

Pop quiz: Where have you seen the symbols shown in Fig. 8.1?

Fig. 8.1 Recycling symbols. PET = polyethylene terephthalate, HDPE = high-density polyethylene, PVC = polyvinyl chloride, LDPE = low-density polyethylene, PP = polypropylene, PS = polystyrene, O = other

This is a book about energy, so you may wonder why it has a chapter about petrochemicals and also why it includes Fig. 8.1 about plastics. Almost all of the symbols in the figure include the letter "P" which you might guess stands for plastic. But actually, the "P" stands for "poly" which means "many." This chapter explains "poly" and what petrochemicals have to do with energy.

Plastics are made by combining compounds, called monomers, into long-chain molecules that contain many of the monomers. The long-chain molecule is called a polymer by chemists. The rest of the world calls it plastic. Figure 8.2 shows how global use of plastics has grown enormously since 1970. Plastic use in 2015 was almost ten times as high as it was in 1970. Look around you now. Almost surely you will see something that is made of plastic. The next highest growth item in the figure is cement, which has grown by almost seven times in the same time period.

© The Author(s), under exclusive license to Springer Nature Switzerland AG 2021   179
M. Cronshaw, *Energy in Perspective*,
https://doi.org/10.1007/978-3-030-63541-1_8

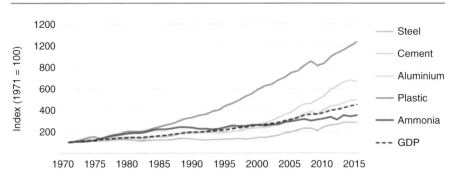

Fig. 8.2 Growth in production

Plastics are a petrochemical. Petrochemicals are products that are made predominantly from oil and/or natural gas. Figure 8.3 shows that petrochemical manufacturing consumed 14% of global crude oil and 8% of global natural gas in 2017.

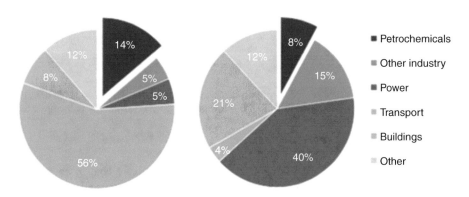

Fig. 8.3 Global demand (2017) for oil (left) and natural gas (right) by sector

The oil and gas that are used to manufacture petrochemicals serve two purposes. The chemicals in oil and gas are transformed into petrochemicals. This aspect is referred to as "feedstock." The other purpose is using the chemical energy in some of the hydrocarbons to enable the transformation from feedstock to petrochemical. The energy used for the transformation is referred to as "process energy." Process energy consists of fuel, steam, and electricity. Figure 8.4 shows that process energy uses about half of the total energy contained in the hydrocarbons that are used for making petrochemicals.

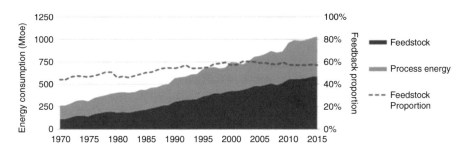

Fig. 8.4 Hydrocarbon use for petrochemicals

There are seven primary petrochemicals: ammonia, methanol, ethylene, propylene, benzene, toluene, and mixed xylenes. About two-thirds of the hydrocarbons used in this sector are for these seven. These seven petrochemicals are key building blocks on which the bulk of the chemical industry is based. Table 8.1 shows the global production of these petrochemicals in 2017.

Table 8.1 Main global petrochemical production

Chemical	Annual global production (2017), million tonnes
Ammonia	185
Methanol	100
Ethylene and propylene	255
Benzene, toluene, xylene	110
Total	650

The largest amount was the combination of ethylene and propylene. Section 2.5.2 explains that ethylene is a hydrocarbon with two carbon atoms and one double bond, and propylene is a hydrocarbon with three carbon atoms and one double bond. The double bond is what enables the polymerization reaction to occur. The corresponding polymers are known as polyethylene (numbers 2 and 4 in Fig. 8.1) and polypropylene (number 5 in Fig. 8.1). Ethylene is also used as a feedstock to make other chemicals including ethylene glycol (the primary component of automobile antifreeze) and polyvinyl chloride (another polymer, number 3, PVC in Fig. 8.1). Propylene is used to make other chemicals also, including isopropyl alcohol (rubbing alcohol) and epoxy resins.

Table 8.1 shows several other chemicals. Ammonia (NH_3) is the starting compound for all nitrogen-containing fertilizers. Agriculture accounts for approximately 80% of total ammonia demand. Ammonia also serves as a chemical building block in several industrial applications including explosives and cleaning products. It may seem strange that ammonia, which contains only nitrogen and hydrogen but no carbon, is called a petrochemical. Also a hydrocarbon feedstock contains carbon, while ammonia does not. The mystery is resolved by noting that the first step in the manufacture of ammonia is the production of hydrogen from methane, LPG, or petroleum naphtha. The carbon contained in those hydrocarbons is removed as carbon dioxide.[1] The hydrogen is then reacted with nitrogen (extracted from air, which is about 79% nitrogen, using cryogenic distillation, similar to that in a natural gas plant) in the Haber-Bosch process to form ammonia.

Another chemical in the table, methanol (CH_3OH), is a liquid that is highly flammable. It is most commonly used to make other chemicals. About 40% of methanol is converted into formaldehyde and further processed into plastics, plywood, paints, explosives, and textiles. It is also used in antifreeze and solvents, and as fuels for some vehicles. It can also be converted to other chemicals.

A key process for producing both ammonia and methanol is steam reforming of natural gas to produce hydrogen. However, oil feedstocks, such as naphtha, liquefied petroleum gas, and fuel oil, can be used instead of methane. China uniquely uses coal as a feedstock for producing both methanol and ammonia.

The final class of petrochemicals in the table consists of benzene (C_6H_6), toluene (C_7H_8), and mixed xylene (C_8H_{10}), collectively referred to as BTX aromatics. These can be modified easily into other chemicals. Their chemical structures are shown in Fig. 8.5. Notice that the three types of xylenes all have the same chemical formula, but they differ in the relative position of the methyl (CH_3) groups on the benzene ring.

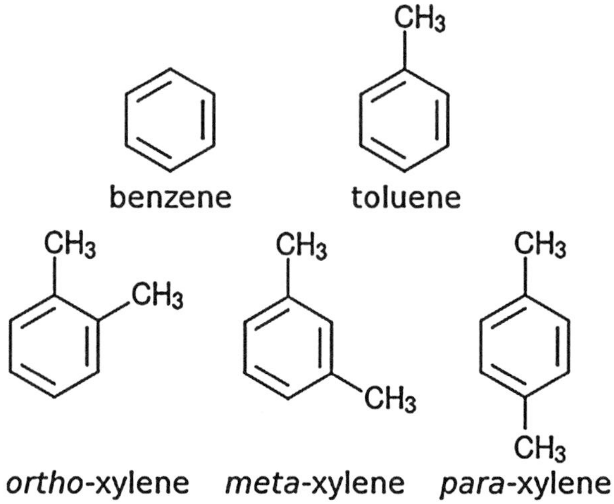

Fig. 8.5 Chemical structure of BTX aromatics

Benzene is used to make several chemicals including polystyrene (number 6 in Fig. 8.1), as well as epoxy resin, nylon, polyurethane (used in varnishes), and alkyl benzene (used in detergents). Toluene is used to make polyurethanes and nylons. Xylene is used to produce other types of plastic including PET (number 1 in Fig. 8.1).

---

[1] This is achieved with two sequential chemical reactions. The formula for the steam-methane reforming reaction is $CH_4 + H_2O = 3 H_2 + CO$. The formula for the water-gas shift reaction is $CO + H_2O = CO_2 + H_2$.

Table 8.1 does not include butylene which is another important petrochemical feedstock. Butylenes have four carbon atoms and either one or two double bonds (see Fig. 2.6 for an example). There are isomers of butylene. One isomer, 1,3-butadiene, has two double bonds. It is polymerized to manufacture elastomers such as polybutadiene and styrene-butadiene rubber (which has largely replaced rubber production from trees), and plastics such as acrylonitrile-butadiene-styrene (ABS). Another isomer, isobutylene, which has one double bond, has been used to manufacture a gasoline additive methyl tert-butyl ether (MTBE). It is also used for copolymerization (polymerization with a combination of monomers) with a low percentage of isoprene (2-methyl-1,3-butadiene) to make butyl rubber.

The feedstocks used to manufacture petrochemicals are produced both in the chemical sector and as by-products from refinery operations. Ethylene, propylene, and BTX aromatics are co-produced in steam crackers (see Fig. 4.18). Ethylene is produced almost exclusively in the chemical sector in steam crackers. Propylene is sourced in large quantities as a by-product of fluid catalytic cracking (FCC) in refineries (see Fig. 4.18). The majority of BTX aromatics are sourced from FCC and continuous catalytic reforming units in refineries. Olefins (hydrocarbons with a double bond) can also be produced from methanol using the methanol-to-olefin process although this is done only in China, where abundant access to coal lowers the cost of producing methanol. Aromatics can also be produced via a similar route, although this process is still in the demonstration phase.

8.2   Uses of Petrochemicals

"Petrochemical" may sound somewhat sinister. But reflect for a moment about the role of plastics, a major petrochemical, in your life. Figure 8.6 may stimulate your reflection. The top right image in the figure may seem out of place to you. Yet it is a polypropylene T-shirt. You may have one of those or perhaps a polyester fleece. (Maybe even polyester pants, but you might not want to admit to that!) You may be shocked to know that polyester is a different manifestation of the PET that is used for the bottles in the top left of the figure!

Figure 8.7 shows the breakdown of global polymer consumption by end-use sector and resin type (PP&A is polyester, polyamide, and acrylic; PUR is polyurethane). You will not be surprised to see how important polymers are for packaging. But you might be surprised to see that textiles are 15% of global polymer consumption, until you realize that nylon, acrylic, and polyester are all polymers.

8.3   Environmental Impacts

There is no doubt that petrochemicals have an impact on the environment. You have probably seen plastic washed up on beaches or riverbanks, and plastic bags stuck in trees.[2] Such adverse impacts can be reduced in several ways. One way is to reuse

---

[2] In the UK, people refer to plastic bags in trees fluttering with the wind as "witch's britches."

Fig. 8.6 Petrochemicals in use

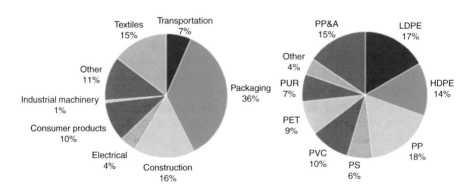

Fig. 8.7 Polymer consumption by end-use sector (left) and resin (right)

products, for example by refilling a purchased bottle of water. Recycling is another method. Recycled plastics can be used for packaging, carpets, and construction materials.

The petrochemical industry is also a source of carbon dioxide emission. The chemical sector emitted about 1.5 gigatons of carbon dioxide globally in 2017, which was 18% of total industrial CO_2 emissions.[3] There are two reasons for carbon dioxide emissions in the chemical sector. First, emissions are caused when fuel is burned to provide energy and steam. This led to about 85% of the emissions. Second, there are emissions due to the difference in carbon content between the feedstock and the product. For example, roughly 0.35 tons of methane feedstock (75% carbon by weight) is required to make one ton of ammonia (0% carbon), so approximately 1.05 tons of carbon dioxide are produced per ton of ammonia as a by-product. Such emissions due to the chemistry are the remaining 15% of carbon dioxide emissions from the sector. Non-CO_2 greenhouse gas emissions from the chemical sector are estimated to be the equivalent of a further 350–400 million tons of carbon dioxide equivalent per year.

8.4   Summary

Hydrocarbons are used to manufacture petrochemicals, including ammonia, methanol, ethylene, propylene, benzene, toluene, and xylene. About 14% of the world's oil production and 8% of the world's gas production are used to make petrochemicals. The components from the oil and gas serve two purposes: they are the feedstock used to create the chemical products, and they provide energy needed for the transformation from feedstock to petrochemical.

Ammonia is widely used as an agricultural fertilizer. Methanol is transformed into many other products such as plastics, plywood, paints, explosives, and textiles. Ethylene and propylene are polymerized into polyethylene and polypropylene. The former is used to make plastic bags and plastic pipes. The latter is used to make plastic containers, furniture, and clothing. One xylene isomer, paraxylene, is the feedstock for making polyethylene terephthalate (PET) which is widely used to make plastic bottles for drinks. Now when you see a triangular recycling symbol with a number in the center, you may think about how hydrocarbons derived from oil and gas were used to manufacture that particular type of plastic.

8.5   Questions

1. The text explains that about 0.35 tons of methane are required to make one ton of ammonia.

   (a) The molecular weights of hydrogen and carbon are 1 and 12, respectively. Show that methane is 75% carbon by weight.
   (b) The molecular weight of oxygen is 16. What is the molecular weight of carbon dioxide (CO_2)?
   (c) Show that about 1.05 tons of carbon dioxide are produced per ton of ammonia.

---

[3] Source: IEA (2018). The Future of Petrochemicals. All rights reserved.

2. The molecular weight of low-density polyethylene (LDPE) is between about 10,000 and 30,000, and the molecular weight range for high-density polyethylene (HDPE) is about 50,000 to 1,500,000.

   (a) What is the range in the number of carbon atoms in LDPE?
   (b) What is the range in the number of carbon atoms in HDPE?

3. More than 480 billion plastic bottles were sold globally in 2016.[4] Assume that they were all made in 2016.

   (a) How many bottles were made per minute?
   (b) The world population was about 7.5 billion people in 2016. How many bottles were sold per person?
   (c) If 1% of the bottles end up in the ocean or rivers, how many bottles ended up there?

4. A plastic grocery bag weighs about 7 g; a paper grocery bag weighs about 55 g.

   (a) How many plastic bags weigh the same amount as one paper bag?
   (b) What are the pros and cons of "paper or plastic"?

5. Look on the Internet to see what polyvinyl chloride is used for.

Appendix: Polymerization

Polyethylene (PE) is the most common plastic in use today. You probably use polyethylene every day. You use it if you get a plastic bag. PE is made from ethylene which has two carbon atoms and is a lightweight gas. By contrast, PE has hundreds or thousands of carbon atoms and is a solid. This appendix gives an introduction to polymerization: the process of combining monomers, like ethylene, to produce polymers like polyethylene.

One method to achieve polymerization is to convert some of the monomers into "free radicals."[5] A chemical bond has two electrons. A free radical is a molecule that had a double bond, but has lost one of its electrons. They are very reactive. If a free radical encounters another monomer, they combine and continue to be a free radical. If an ethylene monomer with two carbon atoms becomes a free radical and encounters another ethylene monomer, then they combine to form another highly reactive free radical with four carbon atoms. If it encounters yet another ethylene monomer, then they combine to form another highly reactive free radical with six carbon atoms.

If one free radical encounters another free radical, then they combine, but the combined compound is no longer a free radical, and is no longer highly reactive. It may encounter another free radical if they are still present in the mixture. In this case they will combine and cease being a free radical. This process continues until all of the free radicals have paired up with other free radicals.

---

[4] https://www.theguardian.com/environment/2017/jun/28/a-million-a-minute-worlds-plastic-bottle-binge-as-dangerous-as-climate-change

[5] The method for creating the free radicals is beyond the scope of this book.

Due to the intrinsic randomness of interactions between the molecules, the resulting polymer at the end of the polymerization reaction is a mixture of molecules with different numbers of carbon atoms. Furthermore, there will be many different geometries for each polymer molecule. Some will have several branches, as well as branches with other branches. Others will have fewer branches.

So there is a distribution of the number of carbon atoms in the molecules in the final polymer mixture. The distribution can be affected by the number of free radicals that are introduced into the monomer mixture initially. The nature of the distribution, as well as the degree to which the polymer is branched or more of a straight chain (see Sect. 2.5.1), determines the mechanical properties of the polymer, such as its density and its mechanical strength. Differences in these properties are reflected in the recycling codes of Fig. 8.1. LDPE is low-density polyethylene, and HDPE is high-density polyethylene. Mechanical properties can also be affected by using a mixture of different monomers, such as butadiene and styrene. The polymerization reaction will be similar to that described above, but there will be a random mixture of the monomers in the polymer.

Mechanical properties can also be adjusted by including additives in the polymer. Polymers are a fascinating topic. Perhaps you will be inspired to study them further in another course.

Sources

Figures

Number	Source
8.2	IEA (2018). The Future of Petrochemicals. All rights reserved
8.3	IEA (2018). The Future of Petrochemicals. All rights reserved
8.4	IEA (2018). The Future of Petrochemicals. All rights reserved
8.5	https://commons.wikimedia.org/wiki/
8.6	https://www.sustainability-times.com/environmental-protection/a-solution-to-dumping-plastic-bottles-a-simple-rubber-band/450,046-39w83z5140jfkuwdjn9dkwB ottles. https://www.keyapparel.com/dryve-tee/dryve-short-sleeve-t-shirt-royal-blue-Polar-King-824-44-front__32431.1582908441. https://www.dw.com/en/eu-reaches-agreement-on-single-use-plastic-ban/a-46797494 46797489_303Plasticware. https://www.njmmanews.com/global-plastic-shopping-bag-market-2020-with-covid-19-after-effects-novolex-advance-polybag-superbag-unistar-plastics-newquantum/ Disposable-Plastic-Bags-Market. https://www.forbes.com/sites/linhanhcat/2019/12/29/plastic-wrap-repels-bacteria/#5d77339a3b14 960x0ClingFilm. https://www.ptonline.com/articles/game-changing-polypropylene-additives-improve-impact-resistance-and-processing-costs-719ptnativemillikendeltamax-collag-500
8.7	IEA (2018). The Future of Petrochemicals. All rights reserved

Table

Number	Source
8.1	IEA (2018). The Future of Petrochemicals. All rights reserved

Energy Industries

9

9.1 Overview

The discussion in this book has been about technologies as well as the supplies and uses of various forms of energy. You have learned about the large variety of uses and technologies. There is plenty that could be added to explore the environmental and economic aspects of energy, but the book is already long enough! There is another non-technological topic that warrants discussion, namely the type of industry structure that is used in different countries and energy sectors. This may sound dry, but it is really important to consider what goals a country might want to achieve with energy industries, and to ponder how those goals can be achieved.

The starting point is to recognize that most countries have some energy resources such as oil, gas, and/or coal within their national jurisdiction (onshore and/or offshore). These resources are valuable only if they are developed, processed, and delivered to end users.

> The country of Bangladesh has natural gas deposits both onshore and off-shore. Some citizens believed that it was not in the national interest to allow investor-owned companies to profit from production of those deposits since they were a national treasure. But without exploration and development they would remain buried treasure. The country has allowed both its national company, PetroBangla, and investor-owned companies to undertake exploration and development.

Table 9.1 lists the criteria that might be considered with regard to energy resources.

© The Author(s), under exclusive license to Springer Nature Switzerland AG 2021
M. Cronshaw, *Energy in Perspective*,
https://doi.org/10.1007/978-3-030-63541-1_9

Table 9.1 Criteria for energy resources	Availability	Environmental impact
	Reliability	Employment
	Affordability	Monitoring
	Profit sharing	Capital requirements
	Expertise	Trade (import and export)

Availability means that the supply of energy should be available for end users. This may sound obvious but access to energy is very limited in many countries. For example, as of 2017 only 9.3% of the approximately 11 million people in the African country of Burundi had access to electricity.[1] Availability is important, but not enough. Even if energy is available to large numbers of people, it may not be reliable. Customers in many countries do not have access to electricity for 24 hours every day. You may have experienced occasional power interruptions. If so, then you know how inconvenient it is. Power interruption is common in some countries. Imagine living without a refrigerator, which is the case for people in the world who do not have a reliable electricity supply.

If energy is available and reliable, then ideally it will also be affordable. Energy that is technically present is not really available if an end user cannot afford to pay for it. But even the three criteria of availability, reliability, and affordability are not enough. Energy may have adverse environmental consequences. We are all familiar with the threat to the global climate caused by emissions of carbon dioxide. Yet, perhaps worse than this is the immediate impact of energy use on health. For example, air pollution in India is estimated to kill 1.5 million people per year.[2] Particulate emissions from vehicles in the Indian city of Delhi are one of the leading causes of pollution. Countries must consider environmental risks and safety with regard to emissions. Risk and safety are intrinsically uncertain by their nature. Yet the possibility of environmental harm, rather than actual definite known harm, is undesirable.

Thousands of people in Ecuador rioted in 2019 after the government ended subsidies on diesel fuel. This caused the price to more than double, severely impacting its affordability.

Table 9.1 shows several other considerations for energy resources. Economies benefit from local employment in the development of energy resources, as well as from their transportation and processing. Such energy activities require inputs such as investment capital and expertise in the relevant areas. Furthermore, imports and exports of energy resources are important. Exports provide income for a country. Alternatively, if a country relies on energy imports, then it is vulnerable to disruption of those imports.

[1] Source: World Bank https://data.worldbank.org/indicator/EG.ELC.ACCS.ZS?locations=BI retrieved Nov. 4, 2019.

[2] https://en.wikipedia.org/wiki/Air_pollution_in_Delhi

All of these considerations need monitoring by government regulatory agencies. These agencies must have the necessary expertise to carry out monitoring. If a country decides to allow investor-owned companies to conduct development, production, transportation, and sales of an energy resource, then it must decide what share of the value of such activities flows to the companies and what share is retained by the country. There are several fiscal tools for sharing the value. Some countries use production sharing contracts that explicitly specify the fractions of production that go to the company and to the country. There are also income taxes, production taxes, and royalties that allow governments to receive a share of the value. In addition, pricing may be regulated. The determination of value sharing is based on two opposing forces. The government wishes to retain a large share, but it must also leave a sufficient amount for the companies so that they earn an acceptable return on their investments. The return that is required by a company depends on the level of risk associated with the energy project. The risk includes geological uncertainty, cost and price uncertainty, and the risk that fiscal terms may be changed by the government in the future.

Countries have been known to appropriate oil and gas assets. In 2007, Venezuela took a majority stake in various oil projects. ExxonMobil and ConocoPhillips quit the country as a result and initiated international compensation arbitration. Shell handed over a controlling interest in the Sakhalin LNG project in Russia in 2006 after a series of governmental environmental challenges. The Russian company Gazprom was the beneficiary of the handover. In 2012 the Government of Argentina appropriated most of the Spanish company Repsol's stake in the company YPF that was developing the giant Vaca Muerta shale gas field.

Given this long list of considerations, it is natural to wonder what type of industry structure is best suited to achieve a good energy outcome for a country. The range of possible industry structures ranges from state- or government-owned enterprises to investor-owned companies. There are examples of each type of firm in different energy sectors. The following subsections provide information for various sectors, as well as suggestions for the underlying logic. Determination of the best industry structure is an active research area in economics. This means that the answer is evolving and changing over time.

9.2 Oil and Gas

9.2.1 Types of Oil and Gas Companies

Chapters 4 and 5 explained that oil and gas operations involve exploration, production, transportation, processing, and sales. Exploration and production are referred to as upstream; oil processing and sales are referred to as downstream; and pipeline transportation and gas processing are referred to as midstream. Some oil and gas companies specialize in only one aspect of oil and gas operations such as upstream. Others have diversified operations including upstream, midstream, and downstream; they are referred to as integrated companies.

Companies choose their focus based on their business strategy. Some companies perceive upstream operations as their strength; others specialize in downstream or midstream. Integrated companies perceive value in selling products such as gasoline, diesel, and jet fuel to end users while also having control over upstream production. For example, ExxonMobil is a multinational integrated oil and gas company. In 2019 it produced about 2.4 million barrels per day of liquids, and 9.4 billion cubic feet per day of natural gas.[3] It refined about 4.0 million barrels per day, i.e., about 1.6 million barrels per day more than it produced, and sold about 5.5 million barrels per day of petroleum products together with about 26.5 million tons of chemical products.

By contrast, Pioneer Natural Resources is an upstream oil and gas company that operates only in the USA. In 2019 it produced 212 thousand barrels of oil per day, 72 thousand barrels of NGLs per day, and 365 million cubic feet of natural gas per day.[4] It does not refine crude oil. Enterprise Products is a midstream company that does not produce oil or gas. It has 50,000 miles of pipelines, 260 million barrels of storage capacity for hydrocarbon liquids, and 14 billion cubic feet of natural gas storage capacity. It also has natural gas processing plants, fractionators, condensate distillation facilities, and docks for marine export of NGLs, polymer-grade propylene, crude oil, and refined products.[5]

9.2.2   National vs. Investor-Owned Oil and Gas Companies

Most countries that have oil and/or gas reserves have state-owned oil and gas companies. They are often referred to as national oil companies (NOCs). A list of several NOCs is shown in Table 9.2. These companies are typically owned by the national government and may be closely linked to a government agency such as a ministry of oil and gas. A government can raise money by selling some or all of its interest in a NOC to investors, which can be a useful source of funds for that government.

In December 2019, the national oil company Saudi Aramco listed 1.5% of its shares for sale on the Tadawul, which is the Saudi stock exchange. The Saudi Government received nearly $26 billion by selling the shares.

---

[3] Production figures are net, i.e., the company's share of the production. Source: ExxonMobil 2019 Annual Report.

[4] Source: Pioneer Natural Resources 2019 Annual Report.

[5] Source: Enterprise Products Partners, Investor Deck, September 2020.

Table 9.2 National oil and
gas companies

Country	Company
Saudi Arabia	Saudi Aramco
Mexico	Pemex
China	China National Petroleum Corporation
Venezuela	Petróleos de Venezuela (PDVSA)
Malaysia	Petronas
Qatar	Qatar Petroleum (QP)
Vietnam	Petrovietnam
Algeria	Sonatrach
Kuwait	KNPC
Bangladesh	PetroBangla
India	ONGC
Angola	Sonangol
Iran	NIOC
Iraq	INOC
Nigeria	NNOC
Libya	LNOC

Oil and gas companies that are not NOCs are referred to as investor-owned companies (IOCs). IOCs are usually corporations that are owned by shareholders. The shareholders may include individuals and/or institutional investors such as pension funds, hedge funds, or private equity companies. Shareholders usually have voting rights that in principle can influence actions by the company, but they do not directly manage the corporations. Instead the companies are run by managers who are employees of the company. The managers, as well as other employees, are usually also shareholders who own a very small percentage of the total number of shares. Even institutional investors typically own only a small percentage of a company's shares. So individual or institutional shareholders cannot dictate decisions made by a company. Despite their small shareholding, some so-called activist large institutional shareholders do influence company behavior. Publicly traded company shares are traded on stock exchanges such as the New York Stock Exchange or the London Stock Exchange. These exchanges allow anyone to buy and sell shares. Consequently, such companies are referred to as public companies. Some companies have shares listed on multiple stock exchanges. This makes it possible for investors to own shares in different currencies, such as US dollars and pound sterling. Not all company shares are traded on such public exchanges. Companies without publicly traded shares are known as private companies. Private companies usually also have shares that determine the value owned by each of their shareholders. Many IOCs have operations in multiple countries. For example, ExxonMobil is a publicly traded integrated company that has operations on six continents.[6] ExxonMobil's shares are traded on the New York Stock Exchange. BP is another integrated public IOC. Its

---

[6]There are seven continents in the world. ExxonMobil has operations in all continents except Antarctica.

shares are traded on the London Stock Exchange and in New York. BP operates in 78 countries. Both ExxonMobil and BP also have renewable energy subsidiaries that complement their oil and gas operations.

> The ownership of energy companies can change over time. Prior to 1986 natural gas in the UK was managed by the government-owned British Gas Corporation. In December 1986 shares in that corporation were offered to investors. At the same time the government created the Office of Gas Supply (Ofgas) to regulate the industry. Ofgas was subsequently merged with the Office of Electricity Generation to form a new regulatory body, Ofgem. A sale to the public is known as privatization. Such privatization provides funds to the government as the public pays money to invest in shares of the company.

Some companies have a hybrid ownership structure, partly owned by a government and partly owned by investors. For example the Norwegian Government owned 67% of the company Equinor as of 2017. Equinor's shares are traded on the Oslo and New York Stock Exchanges. As another example, the Brazilian Government owns 54% of the company Petrobras, while the Brazilian Development Bank and Brazilian Sovereign Wealth Fund[7] each own 5%. The remaining 36% of the company is owned by investors.[8]

NOCs can have two main roles with regard to oil and gas operations. One role is to find and produce oil and gas. They may do so both within the country's territory (onshore and offshore) and in other countries. In addition, NOCs can be involved in refining, processing, and petrochemicals by constructing facilities in their country, participating in joint ventures with other firms in those facilities, and/or investing in such facilities in other countries. The second role of a NOC is to offer opportunities for other companies to conduct oil and gas operations within their country, to monitor those operations, and often to buy oil and gas produced by those companies. Such dual roles can cause difficulties. IOCs may be reluctant to invest in a country that has a NOC, believing that the NOC retains the best opportunities for itself, leaving only second-tier opportunities for others. Despite this potential problem, many countries have various oil and gas operations within their territory, some operated by a NOC and others by IOCs.

The largest producers of oil and gas tend to be NOCs as shown in Table 9.3. Some of the companies in the table are shown as NOC/IOC. Shares of these companies are traded on stock exchanges, but governments also own a substantial number of the shares.

---

[7] A sovereign wealth fund is a government-owned fund that invests in various businesses.
[8] Wikipedia.

Table 9.3 2018 oil and gas production

Company	Type	Daily production million BOE/day
Saudi Aramco	NOC/IOC	13.6
Gazprom	NOC/IOC	10.5
Rosneft	NOC/IOC	5.7
ExxonMobil	IOC	3.8
BP	IOC	3.7
Royal Dutch Shell	IOC	3.6
Chevron	IOC	2.9
Total	IOC	2.7
Eni SpA	NOC/IOC	1.8

BOE is barrels of oil equivalent. Typically 6 MCF of gas is 1 BOE

The company that had the largest production in 2018 was Saudi Aramco, which is a NOC but with a small portion of its shares owned by investors. As discussed in Chap. 4, Saudi Arabia is one of the largest producers of crude oil. Saudi Aramco is responsible for all hydrocarbon production in Saudi Arabia. It is a huge company. Besides oil and gas production, it also has large capacity for refining as well as interests in petrochemicals. ExxonMobil is also a huge company. It was number 2 in the 2019 Fortune 500 listing of companies ranked by revenue.[9] Yet, the table shows that its oil and gas production in 2018 was dwarfed by Saudi Aramco's and Gazprom's production, and was also lower than the production of another Russian company Rosneft.

9.2.3   OPEC

The Organization of Petroleum Exporting Countries (OPEC) was formed at a conference in Baghdad in September 1960. Initially there were five member countries: Iran, Iraq, Kuwait, Saudi Arabia, and Venezuela. The objective of OPEC is to coordinate and unify petroleum policies among member countries. Each of the founding countries had large oil reserves. Up until the formation of OPEC the oil industry had been dominated by large multinational oil companies, nicknamed the "Seven Sisters."[10] These companies controlled about 85% of the world's petroleum reserves. The OPEC member countries wished to exert more control over reserves within their boundaries. They were able to do so by forming an intergovernmental organization.

---

[9] Based on fiscal years that ended on or before January 31, 2019. The Fortune 500 includes companies that are incorporated in the USA, operate in the USA, and file financial statements with a government agency, even if they are private companies. Walmart was number 1.

[10] The Seven Sisters were Anglo-Iranian Oil Company (now BP), Gulf Oil (now part of Chevron), Royal Dutch Shell, Standard Oil Company of California (SoCal, now Chevron), Standard Oil Company of New Jersey (Esso, later Exxon, now part of ExxonMobil), Standard Oil Company of New York (Socony, later Mobil, now part of ExxonMobil), and Texaco (later merged into Chevron).

Additional member countries joined OPEC over time. Some member countries have since withdrawn. As of 2019 there were 14 member countries (see Table 9.4). OPEC has influenced oil prices and production. During times of low price, OPEC implemented a production quota system to limit production in order to increase oil prices. However, member countries did not always honor their quotas. Countries have an incentive to sell more oil in order to increase their revenue.

Table 9.4 OPEC member countries as of 2019	
	Algeria
	Angola
	Congo
	Ecuador
	Equatorial Guinea
	Gabon
	Iran
	Iraq
	Kuwait
	Libya
	Nigeria
	Saudi Arabia
	United Arab Emirates
	Venezuela

Figure 9.1 shows that OPEC member countries have produced a large share of the world's oil: between about 30% and 50% over time. Such a large share means that OPEC has market power over oil prices. However, the extent of this power is somewhat moderated by the fact that there are several autonomous member countries, so OPEC struggles to act as a unilateral body with a single objective.

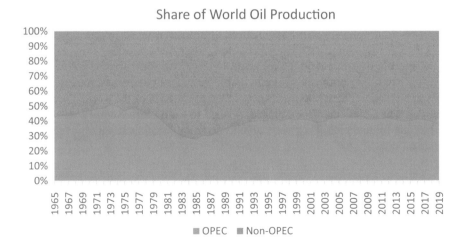

Fig. 9.1 OPEC's share of global oil production

Note how OPEC's share of world oil production dropped dramatically from 1979 to 1985. The 1970s were an extremely turbulent period in oil markets. In 1973 OPEC imposed an oil embargo, limiting exports of crude oil to the USA and the Netherlands. OPEC also implemented a quadrupling of the oil price. In 1979 there was a war between Iran and Iraq which led to another disruption in oil supplies. The oil price doubled again in this period. Given the uncertainty about OPEC oil supplies, oil companies actively increased supplies from non-OPEC countries after 1979. Figure 9.2 shows that total world oil production was almost constant from 1973 to 1985, but OPEC production was cut in half from about 30 million barrels per day to about 16 million barrels per day. The drop in OPEC production was made up by production from non-OPEC countries.

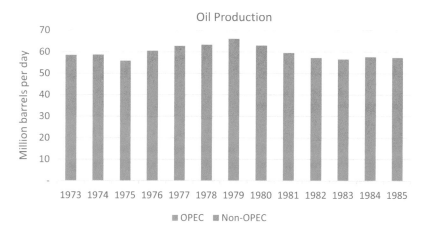

Fig. 9.2 OPEC and non-OPEC oil production

9.3 Coal

The coal industry structure is different in various countries. Some countries have primarily state-owned production, while others have primarily investor-owned production. Some companies mine only coal, while others produce a variety of metals as well as coal. Table 9.5 lists the names of several large mining companies that produce coal. The table shows that none of the companies is headquartered in the USA. Most of these companies produce a variety of metals as well as coal. For example, Glencore produces and processes copper, gold, zinc, lead, nickel, ferroalloy, aluminum, and iron ore as well as coal. Large mining companies tend to be multinational with operations in several countries. Their shares trade on many different stock exchanges in the world.

Table 9.5 Large mining companies with coal production

Company	Mining revenue $billion (2017)[a]	Headquarters
Glencore	$90	Switzerland
Rio Tinto	$40	UK and Australia
BHP	$34	Australia
Vale	$34	Brazil
China Shenhua Energy Company	$30	China
Anglo American	$26	UK

[a]https://www.mining-technology.com/features/worlds-biggest-mining-companies-2018/

There are also state-owned coal companies. For example, Coal India Limited is owned by the Indian Government. It is a huge company. It produced 625 million short tons of coal in its 2017/2018 fiscal year.[11] The whole country of India produced a total of 789 million short tons in 2017,[12] so Coal India's production was about 79% of the country's total.

In contrast to India with its large government share of production, investor-owned companies produce most of the coal production in the USA. The USA produced 1000 million short tons of coal in 2014.[13] Table 9.6 shows that the top three US coal companies produced 411.1 short tons in that year, i.e., 41% of the US production. The top ten coal companies produced about 73% of the US total.[14] When a small number of companies produce a large fraction of the total they can have market power, i.e., they can strongly influence the price. However, it is important to consider the relevant market when assessing market power. In the case of coal, there are two reasons why the relevant market is not simply coal sales in the USA. Firstly, there are large amounts of coal exports from the USA, so the relevant market is larger than the US coal market alone. In addition, coal competes with natural gas as a fuel for electric power plants, and fossil fuel power plants compete with other electricity generation technologies. The price of natural gas in the USA has been low for several years, since fracking has greatly increased natural gas production. In fact, the low price of natural gas has created financial difficulties for US coal companies. Many of them have declared bankruptcy recently due to price competition from natural gas.

Table 9.6 Top US coal producers (2014)

Company	Million short tons
Peabody Energy	189.5
Arch Coal	135.8
Cloud Peak Energy	85.8

---

[11] 567 million metric tons from https://www.coalindia.in/en-us/company/aboutus.aspx, retrieved Nov. 1, 2019.

[12] BP Statistical Review, 2018, shows production of 716 million tonnes.

[13] Wikipedia.

[14] Ibid.

9.4   Electricity

Just as with oil, gas, and coal there are both national and investor-owned electricity companies. There are also companies whose shares are traded on a stock exchange with the government owning some of those shares. For example, Public Power Corporation is the largest electric power company in Greece. Shares of the company are traded on the Athens Stock Exchange, and the Greek Government owned 51.1% of the company's shares as of 2018.[15] Other countries with national electricity companies include Guinea, Jordan, Oman, Bulgaria, and Senegal. There are also electricity companies that are not owned by federal governments, but rather by local communities. Such companies are known as municipal utilities.

As explained in Chap. 7, there are three sectors of the electricity industry: generation of power, transmission of power from generators to market centers, and distribution of power within market centers. There must be coordination between these sectors to promote reliability of supply. Since generation, transmission, and distribution may be owned by different entities, such coordination can be challenging. Different countries and different regions within countries have taken different approaches to the coordination.

One key issue is monopoly power. In many situations it is most cost effective to have a single facility providing power, for example just one power line to a home or one transmission line from a generation facility to a market center. This is referred to as a natural monopoly. Without regulation, the owner of a natural monopoly could charge a high price for its service. In order to avoid such high prices, there are public utility commissions (PUCs) that regulate pricing. PUCs ensure pricing levels that are high enough for companies to have an economic incentive to provide services but no higher so that consumers do not pay excessively high prices. There is a rich literature about such regulation.

9.5   Petrochemical Companies

Table 9.7 lists the top ten global petrochemical companies based on 2008 revenues.

Table 9.7 Top ten petrochemical companies

Company	Headquarters	2008 revenues $ (billion)
BASF	Germany	62.30
Dow Chemical	USA	57.51
ExxonMobil Chemical	USA	55
LyondellBasell Industries	Netherlands	51
INEOS	UK	47
SABIC	Saudi Arabia	37.66
Formosa Plastics Corporation	Taiwan	31.5
Sumitomo Chemical	Japan	16.65
DuPont	USA	31.83
Chevron Phillips	USA	13

---

[15] Wikipedia.

These companies have a global presence. For example, BASF operates in Europe, Asia, Australia, the Americas, and Africa. It produces solvents, amines, resins, glues, industrial gases, petrochemicals, inorganic chemicals, plastics, urethanes (used in foams), coatings, catalysts, fungicides, herbicides, insecticides, and bio-technological products.[16]

Most of these companies are publicly traded or are subsidiaries of publicly traded companies. Although SABIC is publicly traded, the Saudi Arabian Government owns 70% of the shares. Chevron Phillips is a company that is jointly owned by Chevron and Phillips 66, which is not affiliated with the oil and gas company ConocoPhillips.[17]

9.6   Summary

Countries with energy resources such as oil, gas, and coal consider several factors when determining access to these resources. It is important that the resources be made available to consumers, and that the supplies be reliable and affordable. Countries must also consider the environmental impacts of resource development and consumption. The economic impacts of such resources can have a big impact, by generating revenue from exports, profit sharing with producers, employment, and taxes.

Many countries have national oil companies (NOCs) that conduct oil and gas activity. Most also offer development opportunities to investor-owned companies. Investor-owned companies (IOCs) have many shareholders, in some cases with the government being a shareholder. Shareholders have only minor influence on company operations and decision-making. Employees of IOCs are responsible for operations and decision-making. They are often shareholders, but with small ownership percentages. IOCs often operate in many different countries. The world's largest oil companies are NOCs, including Saudi Aramco and Gazprom. Oil and gas companies operate in the upstream, midstream, or downstream sectors. Some specialize in one of these, while others, known as integrated companies, operate in all of them.

OPEC is an intergovernmental organization that was formed in 1960. It had 14 member countries as of 2019. OPEC coordinates oil production among its member countries with a goal of supporting the oil price. OPEC countries produced between 30% and 50% of the world's oil from 1965 to 2019.

Most coal around the world is produced by mining companies that produce multiple commodities besides coal, including copper, gold, zinc, lead, nickel, and iron ore. Electricity companies in some countries are government owned, either by the central government or as municipal-owned operations. Investor-owned electric utilities are also common. It is often cost effective for there to be one electricity operator to avoid duplication of facilities. Such operators are regulated by public

---

[16] https://en.wikipedia.org/wiki/BASF

[17] Phillips 66 was separated from ConocoPhillips in 2012. https://www.phillips66.com/about

utility commissions to ensure that prices are reasonable for customers but also high enough to support investment.

Large petrochemical companies operate in many countries, as do many oil and gas firms. Their expertise and proprietary technologies can be implemented in different locations.

9.7 Questions

1. *This question presents a highly stylized version of a production sharing contract such as used by some countries when awarding oil and gas exploration and development rights to an investor-owned company.* Suppose that the value created by finding, developing, and producing an oil and gas resource is $50 billion, and that the total cost to find, develop, and produce the resource is $10 billion. If the company is allowed to recoup the cost and 20% of the remaining value after costs, what share of the total value does the country receive?
2. In 2019 ExxonMobil had net production of 2.386 million barrels of liquids and 9.394 billion cubic feet of natural gas per day.

   (a) How many million BOE per day did it produce? Convert gas to BOE with a factor of 6 MCF to 1 BOE.
   (b) What percentage of this amount was liquids?

3. Production by Saudi Aramco in 2019 is shown in the table below.

   (a) How many BOE per day did the company produce in 2019? Convert natural gas and ethane to BOE using factors of 5.4 and 3.33 MMSCF per BOE, respectively.
   (b) What percentage of this amount was liquids?
   (c) Compare the production of ExxonMobil and Saudi Aramco.

2019 Production	MBPD	MMSCFD
Crude oil (with blended condensate)	9,943	
Condensate	202	
Propane	535	
Butane	319	
Natural gasoline	222	
Subtotal	11,221	
Natural gas		8,978
Ethane		960

4. As of year-end 2019, ExxonMobil had proved liquid reserves of 14,598 million barrels and proved natural gas reserves of 47,080 BCF. Saudi Aramco had proved crude oil and condensate reserves of 201,907 million barrels and proved natural gas reserves of 190,575 BCF.

   (a) What were the proved reserves in BOE for the two companies? Use a conversion factor of 6 MCF per BOE.
   (b) What percentage of the BOE reserves was liquids for the two companies?
   (c) How do the reserves of Saudi Aramco compare with those of ExxonMobil?

5. Total 2019 world production of crude oil (including shale oil, oil sands, conden-
   sates, and NGLs) was 95.1 million BPD; US production was 17.0 million BPD;
   and Saudi production was 11.8 million barrels per day.

   (a) What percentage of the global total was produced in the USA?
   (b) What percentage of the total was produced in Saudi Arabia?
   (c) What percentage of the total was produced by Saudi Aramco?
   (d) What percentage of the total was produced by ExxonMobil?

6. In 2019 ExxonMobil's net income after tax (i.e., profit in $million) by business
   segment was as follows: upstream $14,442, downstream $2,323, chemicals
   $592, and corporate and financing –$3,017. In the same year, the chemical company BASF had net income after tax of 8,421 million euros. The net income is
   reported in different currencies due to the different headquarter locations.

   (a) What was the total net income for ExxonMobil?
   (b) Excluding corporate and financing, what proportion of net income after tax
       was contributed by each of ExxonMobil's business segments?
   (c) The average exchange rate in 2019 was $1.12 per euro. Compare the total
       income of the two companies.
   (d) How does BASF's net income after tax compare ExxonMobil's net income
       from chemicals? Neglect ExxonMobil's negative net income from corporate
       and financing.

Sources

Figure

Number	Source
9.2	BP Statistical Review of World Energy 2019

Tables

Number	Source
9.3	Economist Magazine, Nov. 2, 2019
9.6	Wikipedia
9.7	Wikipedia

Technological Change

10

10.1 History as a Guide to the Future

You have learned a great deal about each of the energy sectors: oil, gas, coal, electricity, and petrochemicals. You have also learned how energy is used in residential, commercial, and industrial settings. Cross-country comparisons shed light on how energy use varies widely around the world, in terms of both energy use per capita and different technologies used for energy supply. A natural question is to wonder how the energy picture will unfold in the future. You are not alone if you are interested in this question. Many government agencies, energy firms, international bodies like the International Energy Agency and OPEC, and consulting firms devote considerable effort to addressing it.

It may seem trite to say this, but it is true: the future will either be the same as the past or be different. There are so many possible different futures and, by contrast, there is only one way for it to stay the same. So it would be very surprising if the future turns out to be the same as the past. There are various forces that will influence future change, including the desire to reduce the environmental impacts of energy, and the remarkable improvements in renewable energy technologies and batteries. This subsection reviews changes that have happened in the past, which may serve as a guide to possible developments in the future.

10.1.1 The Long View Back

Figure 10.1[1] shows a long-term history of energy use in the USA. Prior to 1850, wood, which is a form of biomass, was the only source of energy in the USA. If you have been camping you know how useful wood is for providing heat. You may even have cooked on a wood fire. In the early nineteenth century and before, wood was virtually the only fuel used for both heating and cooking.

---

[1] Biomass includes wood and corn (for liquid fuels).

© The Author(s), under exclusive license to Springer Nature Switzerland AG 2021
M. Cronshaw, *Energy in Perspective*,
https://doi.org/10.1007/978-3-030-63541-1_10

Energy consumption in the United States (1776-2040)

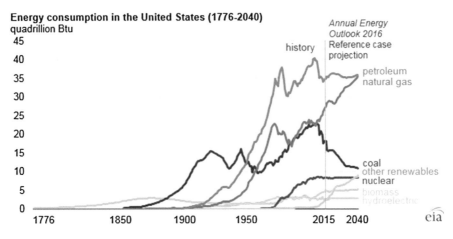

Fig. 10.1   History of US energy consumption

A shift towards coal began in about 1850, although energy supply from wood continued to increase until about 1870. It takes a while for a new technology to displace an existing one. Coal is a more convenient fuel than wood. It has twice the energy density (BTU per pound) as wood. If you have ever hauled wood or coal you will appreciate the benefit of carrying twice as much energy in one load. The second half of the nineteenth century was a time of rapid industrialization in the world and in the USA. The figure shows that energy consumption from coal in the early 1900s dwarfed that from wood at its peak by about four times. Energy played a key role in the industrialization. Energy consumption from coal grew substantially until about 1920, when it declined, but then recovered for several more years. Figure 10.2 shows a behemoth coal-fired train that was used in the USA starting in the 1940s. The tender behind the locomotive (with the Union Pacific logo) carried about 25 tons[2] of coal for fuel.[3]

Starting in about 1920 petroleum and natural gas consumption became significant. Oil was discovered in the USA much earlier, in 1859 in Pennsylvania.[4] But it took some time for its use to become widespread. The first use of oil was to provide kerosene for lamps. This was cheaper and more convenient than candles or the whale oil that had been commonly used. However, electricity became a major challenge to kerosene. Electric lighting was much better than kerosene lamps: no soot and no lamp filling were required. Thomas Edison's company, the Edison Electric Illuminating Company of New York, brought electric lighting to parts of Manhattan

---

[2] The 25 tons of fuel is miniscule compared to the 15,000 tons pulled by unit trains such as that shown on Figure 6.10.

[3] http://www.northeast.railfan.net/challenger.html

[4] There are two excellent books about the history of energy: "The Prize" by Daniel Yergin and "Energy and Civilization" by Vaclav Smil.

Fig. 10.2 Coal-fired train. Union Pacific 4-6-6-4 Legacy Challenger

in 1882.[5] Adoption of electric lighting was slow. By 1925 only half of US homes had access to electricity.[6]

It was the development of another technology that led to the rapid growth in the use of oil, preventing its demise due to the competition from electricity. Henry Ford built an experimental car in 1896. The Model T Ford was introduced in the USA in 1908.[7] Cars were a great improvement over horse-drawn vehicles. The adoption of cars led to large demand for motor fuel. Petroleum was the source of this fuel.

Natural gas was used for lighting streets and in houses in the eighteenth and nineteenth centuries.[8] But very little energy is required to provide light. It was other uses that led to an increase in the use of natural gas. US consumption of natural gas increased substantially in the twentieth century when pipelines were built to bring natural gas to homes for heating and cooking, to industry for process heat, and to power plants for electricity generation.

Nuclear energy was developed during the Second World War, in the form of nuclear bombs that were dropped on Hiroshima and Nagasaki, Japan. The enormous energy provided by nuclear reactions was harnessed after the war in the 1950s for use in naval applications, particularly submarines. This led to the use of nuclear reactors for generating electricity. A demonstration 60 MW pressurized water nuclear reactor started in Pennsylvania in 1957.[9] Figure 10.1 shows the rapid increase in the use of nuclear energy after that. Growth in the use of electrical

---

[5] https://www.nps.gov/edis/learn/kidsyouth/the-electric-light-system-phonograph-motion-pictures.htm

[6] Ibid

[7] https://www.britannica.com/topic/Ford-Motor-Company

[8] https://www.apga.org/apgamainsite/aboutus/facts/history-of-natural-gas

[9] https://www.world-nuclear.org/information-library/current-and-future-generation/outline-history-of-nuclear-energy.aspx

nuclear power came to a halt in about 2012, due to cumulative effects of three major nuclear accidents. In 1979 there was an accident at Three Mile Island in Pennsylvania; in 1986 there was a meltdown at Chernobyl, Ukraine; and in 2011 there was a meltdown after a tsunami at Fukushima, Japan.

There was something of a resurgence in the use of biomass in the USA after 1970. In the USA in 2017 47% of energy from biomass was from biofuels (mainly ethanol from corn that is blended into gasoline), 44% was from wood and wood-derived biomass, and 10% was from methane produced from landfills.[10]

The preceding discussion summarized long-term historical trends in US energy consumption. There were also short-term fluctuations. Figure 10.1 shows that petroleum consumption dropped in the late 1970s and early 1980s and again after 2008. The 1970s' drop was in response to the large price increases initiated by OPEC (see Sect. 4.8). The 1980s' drop was in response to a large decrease in the price of oil. The price of the benchmark West Texas Intermediate (WTI) crude oil dropped from about $140 per barrel in July 2008 to about $30 per barrel in December 2008, just 5 months later. This caused a huge reduction in upstream oil and gas activity, since new investments became economically unattractive.

The figure also shows that natural gas consumption declined during the 1970s. This was largely due to US regulation. The price for gas delivered across state lines became regulated then, and the 1978 Fuel Use Act not only prohibited new power generation by natural gas, but also set a time table for shutting down existing gas-fired power plants, due to a perception that natural gas was a scarce resource.[11] How the times have changed! Now there is a glut of gas in the USA, and power generation with gas is favored over coal!

There have been three periods of decline in the US consumption of coal: in the 1920s, the 1940s, and recently. The decline in the 1920s was due to an economic recession after the First World War.[12] The decline in the 1940s occurred as petroleum and natural gas were substituted for coal.[13] Recently coal use has declined from its peak of about 23 quads due to (1) concern about its large emissions of carbon dioxide and (2) competition from natural gas for power generation, as a result of the extraordinary increase in US natural gas production brought about by horizontal drilling and fracking.

In early 2020 the covid-19 virus spread across the world. By August 2020 there were 20 million cases worldwide and about 750,000 deaths. This highly infectious disease led to large changes in behavior in almost every country. Large gatherings were prohibited; economic activity faltered as demand for goods and services fell; large numbers of employees were made redundant, so the unemployment rate rose precipitously leading to a reduction in demand for many goods. Transportation also declined as fewer people were commuting to work. Also, international air

---

[10] Rounding affects the numbers. https://www.eia.gov/energyexplained/biomass/

[11] "U.S. Energy: Overview and Key Statistics," Congressional Research Service, June 27, 2014 p.21

[12] "Annual Outlook for US Coal: Volume 985" May 1985, EIA retrieved online

[13] Ibid

travel was restricted as countries moved to protect themselves from infection by people arriving from outside their borders. The impact on energy sectors has been unprecedented. Demand for energy fell by 18–25%.[14] Road passenger traffic fell by about 70% in China, about 80% in India, and almost 50% in the USA.[15] Weekly electricity demand dropped by about 20% in several countries.[16] Energy companies are slashing their expenditures for new projects in response to the reduction in demand for energy. Many of these impacts will likely be corrected when the virus has been controlled. However, some impacts may persist. For example, people may continue to work from home, which will reduce the amount of road and air travel.

There are a few things to learn from this historical review:

1. Energy consumption in the USA has generally increased over time, but with some short-term declines over periods of a few years.
2. New energy technologies have been introduced over time.
3. There is a lag in the widespread adoption of new types of energy.
4. New technologies can displace older technologies.
5. Environmental and regulatory issues are important determinants of energy use.
6. Extraordinary events such as the global covid-19 pandemic can have a large impact on energy use.

The next two subsections provide details of two recent energy transitions that occurred: one in Germany, the other in Japan.

10.1.2 Energiewende: A Recent Transition[17]

In late 2010 Germany passed legislation to radically change its energy system. The legislation is referred to as the Energiewende, which means energy transition. Germany committed to reduce its emissions of greenhouse gasses (GHG) by 80–95% (relative to 1990 levels) by the year 2050. It also planned for renewable energy to provide 60% of its energy by 2050. By 2014, Germany had reduced its emissions of GHG by 27% relative to 1990. However, it will not achieve an intermediate goal of 40% reduction in GHG emissions by 2020. Energiewende anticipated a more active market in carbon trading than has materialized. Carbon trading makes it more expensive to use energy technologies with high emissions of carbon dioxide, which enhances the cost-effectiveness of renewable wind and solar technologies.

Despite the limited use of carbon trading, Germany has reduced its use of coal for electricity generation. Figure 10.3 shows that the share of coal in Germany's energy consumption dropped from 37% of total energy consumption in 1990 to 18% in 2019.

---

[14] https://www.iea.org/topics/covid-19

[15] Ibid

[16] Ibid

[17] This subsection relies on information at https://en.wikipedia.org/wiki/Energiewende.

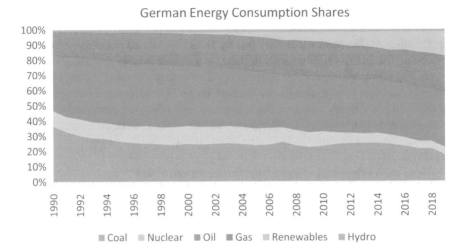

Fig. 10.3   German energy consumption shares by type

Additional legislation that passed in the year after the Energiewende also called for nuclear energy to be phased out by 2022. Most German nuclear power plants have been shut down; the share of nuclear in German energy consumption dropped from 10% in 1990 to 5% in 2019. By design renewable energy has increased in importance. Its share increased from about 0.1% in 1990 to 16% in 2019.

Energiewende also called for improvements in energy efficiency. Figure 10.4 shows that Germany has achieved reductions in total energy consumption as well as energy consumption per capita, which illustrates that there have been improvements in energy efficiency. However it is notable that substantial improvements in each of these measures occurred prior to the passage of the Energiewende legislation in 2010.

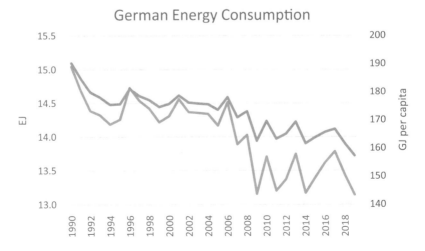

Fig. 10.4   German energy consumption

It was an ambitious agenda to phase out both coal and nuclear. As of 1990, these two energy sources provided almost 50% of German energy consumption. The agenda has presented several challenges. Large-scale generation of renewable energy by solar and wind requires investments in new locations and therefore also new investments in electricity transmission lines. There has been public opposition to the impact of the new transmission lines that would be required. Aboveground transmission lines are unsightly as are large-scale wind and solar farms.

The cost of energy to German consumers is higher than that for consumers in many other European countries. The imposition of constraints in any form on any system results in higher costs. Offsetting this, there are surely environmental benefits. Public opinion, which initially was hugely in favor of the anticipated environmental benefits, has become somewhat less enthusiastic. In addition, Germany will become more reliant on imports of natural gas from Russia. Natural gas is a clean burning fuel, but it makes Germany increasingly reliant on cooperation from another country with associated geopolitical implications.

10.1.3 Fukushima, Japan[18]

There was a huge magnitude 9 earthquake off the coast of Japan on March 11, 2011. The earthquake caused a 45-foot-high tsunami causing failures in backup generators that were providing power to cooling water pumps at the Fukushima Daiichi nuclear power plant in Japan. The loss of cooling caused three nuclear meltdowns, three hydrogen explosions, and a large release of radioactive contamination. At the time, Japan had 54 operating nuclear reactors that produced about a third of Japan's electricity. Figure 10.5 shows that nuclear energy provided about 13% of Japan's total energy consumption in 2010, the year before the accident.

All 50 intact nuclear reactors in Japan were shut down after the accident. Figure 10.5 shows that nuclear power contributed hardly anything to Japan's energy consumption mix from 2012 to 2016. In 2012 the Japanese Government announced their intention to phase out all nuclear power by 2040, when existing plants would reach the end of their 40-year operating lives. However, some nuclear plants have resumed operations since 2015.

Figure 10.6 shows that both total Japanese energy consumption and per capital energy have fallen since 2010.[19] Increased use of oil, gas, and coal made up for some of the energy lost due to the nuclear shutdowns. Fossil fuels (oil, gas, and coal) together with nuclear provided 20.0 EJ of energy in 2010. In 2012 this combination provided 18.9 EJ, a drop of about 1.1 EJ or 5% in two years. Consumption from

---

[18]This subsection relies on information at https://en.wikipedia.org/wiki/Fukushima_Daiichi_nuclear_disaster and https://www.scientificamerican.com/article/6-years-after-fukushima-japans-energy-plans-remain-murky1/#:~:text=When%20the%20earthquake%20and%20tsunami,intact%20reactors%20one%20by%20one.&text=The%20plan%20did%20not%20rule%20out%20building%20new%20nuclear%20plants.

[19]Per capita consumption in Japan has been similar to that in Germany, but slightly lower.

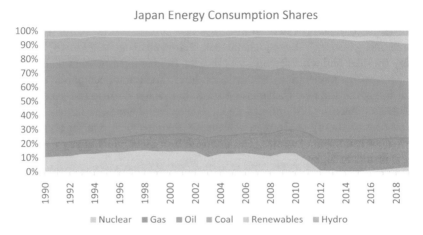

Fig. 10.5   Japanese energy consumption shares

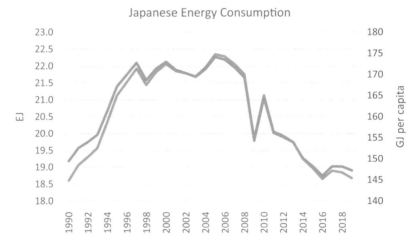

Fig. 10.6   Japanese energy consumption

nuclear energy dropped by 2.6 EJ from 2010 to 2012, but energy from fossil fuels rose by 1.4 EJ.[20]

Figure 10.6 also shows a big drop in total and per capita consumption from 2008 to 2009. Figure 10.4 shows that similar drops occurred in Germany. There was a global financial crisis in 2008 that had a profound impact on economic activity in many countries including the USA. The crisis resulted in decreases in energy supply and demand. The impact was substantial. The price of West Texas Intermediate crude oil at Cushing Oklahoma fell from a high of about $140 per barrel in July

[20] Rounding means that the sum of the fossil fuel increase (1.4 EJ) and the drop in the combination (1.1 EJ) is not precisely equal to the drop in nuclear (2.6 EJ).

2008 to about $30 per barrel by the end of that year, almost an 80% drop in the price over a period of 5 months. Correspondingly, oil and gas companies reduced their investments in new projects.

Japanese consumers responded to the nuclear shutdowns with strong energy conservation efforts and increased energy efficiency. There were no shortages of electricity. The price of electricity rose by more than 20% in 2012 and 2013 due to the higher cost of generation by fossil fuels compared to nuclear. But the electricity price stabilized afterwards.

In summary, both in Germany and in Japan disruptions in the energy supply system (one by regulation, and the other by accident) did not result in major disruptions in energy consumption. In both cases, energy conservation and fuel switching covered the impacts of the disruptions. However, the price of energy rose in both locations.

10.2   100% Renewable?

You may come across the topic of "100% renewable" in your conversations or reading. Isn't that obviously a good thing? The alternative of using up a nonrenewable resource so that it is not available for others or future generations sounds irresponsible, uncaring, and certainly not sustainable. Perhaps there is nothing interesting to say about "100% renewable" since it is obviously the right answer. But this book provides some details that will help you understand that there is something interesting to say about 100% renewable. Solar, wind, and hydroelectric are renewable forms of energy. How would our energy system perform using only these three sources of energy? Well, recognize that each of these technologies generates electricity, a very convenient source of energy. Solar can also generate some heat for homes and their hot water. But there are other important considerations. Solar and wind are intermittent energy sources. If you need energy when they are not being generated, what will you do? There is not enough hydropower to make up for the loss. A seemingly obvious answer is store energy when solar and wind are generating and to make the stored energy available when it is needed. In the case of electricity, this generally means having batteries. As of today, no battery technologies exist that can store enough grid energy that will serve millions of customers for a long enough duration. The technology is advancing extraordinarily rapidly, so the future may be quite different. But it will take some years (or decades) for the technology to make the enormous improvements that are required.

Even then, there is the issue of energy for transportation. There is certainly a significant increase in the number of electric cars on the roads. Those cars and their built-in technologies have solved the energy storage problem. Cars have the advantage that they are supported by gravity. So the weight of batteries is a surmountable concern. Airplanes lack the support of gravity. To the contrary, energy is required to overcome the force of gravity trying to bring the plane to earth. The heavy weight of batteries means that a significant portion of the energy stored in the batteries would be used solely to keep them aloft. An interesting technological feat, but not at all aligned with the goals of air transport, i.e., carrying people and cargo.

Renewable energy has considerable environmental benefits. It should be *part* of the energy mix. But a lot of details need to be worked out before it can become our sole source of energy.

10.3   A Possible Future

If only there was a crystal ball that accurately predicted the future! This section makes an attempt at a forecast, although forecasts are notoriously unreliable:

1. The demand for energy is likely to increase over the long run as global economic development continues. The increase in demand may be moderated by efficiency measures, such as more insulation in buildings, better fuel economy in transportation (more miles per gallon), and more use of electric and hybrid vehicles that use energy more efficiently than do internal combustion vehicles.
2. There is likely to be moderation in the use of coal due to the large emissions of carbon dioxide that are associated with its combustion.
3. The shift away from fossil fuel-powered internal combustion towards electric vehicles has the potential to reduce carbon emissions, provided that the energy used to generate the electricity for charging vehicles does not result in a large increase in emissions.
4. Electricity use is likely to increase since it can be generated with no carbon emissions by solar, wind, and nuclear technologies.
5. Research will continue on solutions for the storage of nuclear waste, which is so long-lived. There may be progress on alternative nuclear technologies such as thorium reactors that are intrinsically safer and whose waste products are less troublesome.
6. Small communities in remote locations will increasingly be served by microgrids that are not connected to a large-scale electrical grid. Electricity supply for microgrids will come from a combination of solar, wind, diesel, and batteries.
7. Use of renewables including solar energy and wind turbines will increase. However there is likely to be opposition to massive increases in the deployment of solar and wind due to the geographic footprint that will be required and the associated impact on view sheds. There has already been resistance to offshore wind turbines in the northeastern USA. Also large-scale increases in wind energy may be limited by the cost of new electricity transmission lines that will be needed to bring the power to demand centers.
8. The intermittent nature of renewables and the time-dependent nature of energy demand will stimulate research on improved energy storage technologies, such as batteries or pumped storage.
9. Hydrogen may be developed as an energy technology, both for transportation and for energy storage. When hydrogen is burned with air, the only emissions are water and nitrogen which is present in the air that is used for combustion.

Currently fossil fuels are used as feedstock to produce hydrogen, and carbon in the form of carbon dioxide is a by-product of this process. However, hydrogen can be produced from water using electricity. There are no associated emissions of carbon dioxide during such operations if the source of electricity is carbon free, such as wind, solar, or nuclear. Use of hydrogen for transportation will require a substantial investment in infrastructure such as filling stations and pipelines to bring the hydrogen to markets. Hydrogen fuel cells that do not burn the hydrogen, but generate electricity directly, may become widely adopted for vehicles.

10. There may be increased use of wood for electricity generation. During the process of photosynthesis trees sequester carbon captured from carbon dioxide in the air as they grow. Thus the net carbon impact of energy from trees is small, since the carbon dioxide emitted during combustion was sequestered during growth.

11. There may be substantial increases in the capture and storage of carbon dioxide from stationary sources such as fossil fuel power plants. A technical solution will be required for dealing with the large amounts of nitrogen that are present in combustion gases when air is used for combustion. Scrubber technology can be used to extract carbon dioxide from combustion gases. However, scrubbers are costly to install and require energy to operate. Another possible solution is to separate oxygen from air before combustion. Then the combustion gases will not contain nitrogen. Oxygen separation from air is currently done on a large scale using cryogenic distillation similar to that used in natural gas plants. However, it is costly.

12. Carbon dioxide is unlikely to be removed directly from the air due to technological limitations, even though there is current work on this.

13. The world will not run out of fossil fuels for a very long time, even though there is a finite supply of oil and gas. The price of these fossil fuels will increase as they become scarce. High prices will encourage efficiency and discourage wasteful use. Air transportation may become a luxury since electrification of large-capacity airplanes is unlikely to occur due to the weight of batteries.

14. This book has not addressed energy economics, but energy prices have been very volatile. Price volatility is likely to continue into the future. At the end April 2020 the price of WTI at Cushing, Oklahoma, was negative! That is, buyers were paid to take oil away from Cushing. This was a very unusual example of price volatility. The negative price arose due to a lack of oil storage capacity, so oil became a waste product for a period of time, not unlike household garbage!

15. The covid-19 pandemic has had a profound impact on energy sectors across the world. Travel is much reduced, so energy demand is substantially lower. The price of energy has fallen a lot in response. Also investment in new energy projects has curtailed, and energy company employees have been laid off in large numbers. Global pandemics may occur again in the future. If so, energy will once again be profoundly affected.

10.4   Summary

History can provide a guide to what may occur in the future. Unsurprisingly the main energy source has changed over the past 200 years. Initially wood was the main source of energy. It provided heat for homes and was used for heating. Wood was gradually replaced by coal, which has a higher energy density (BTU per pound). Coal not only replaced wood, but also led to a large increase in energy use, as industry grew enabled by technologies that converted heat energy into work. The transition was slow. It took about 40 years for the energy from coal to reach the same level as that from wood in the USA.

The discovery of oil and construction of pipelines for both oil and natural gas led to yet another big increase in energy consumption in the USA. However, this increase also occurred slowly, over a period of decades. Nuclear energy for electricity generation became more widespread starting in the 1950s. Hydroelectricity grew in importance in the mid-twentieth century. Renewable energy from solar and wind is becoming increasingly important but it still provides a relatively small amount of energy. They are tending to replace coal as an energy source since they do not emit carbon dioxide during operations, unlike coal which has large emissions per unit of energy produced.

Both Germany and Japan have undergone energy transitions in the twenty-first century, phasing out nuclear power (and coal in the case of Germany). Germany chose its transition in order to reduce the environmental impacts of coal. Japan responded to the nuclear accident at Fukushima. There have not been energy shortages in either country. Both countries responded by improving energy efficiency and reducing per capita consumption. However, energy prices increase in response to transitions.

Predictions about the future are notoriously unreliable. The recent covid-19 pandemic has had a big impact on energy demand and makes predictions even more uncertain. But over the long run, energy demand is likely to increase together with economic development. But energy demand may increase more slowly than the rate of economic development as energy is used more efficiently. Electricity is likely to become an increasingly important source of energy, since there are electricity technologies (solar, wind, and nuclear) that do not emit carbon dioxide during operations. The intermittent nature of solar and wind provides a strong incentive for the development of better electricity storage technologies. Battery technology has improved substantially, but more improvements are needed.

10.5   Questions

Chapter 1 closed with a list of questions. They are listed below again. Enjoy discussing them. Have your thoughts changed now that you have finished this book?
1. Will we run out of oil?
2. Will there be an environmental catastrophe due to carbon dioxide?
3. Is fracking bad?

4. Should you turn off all electrical appliances when not in use?
5. What energy conservation measures make sense?
6. What should the role of renewable energy be?
7. Do energy companies have market power?
8. What energy problems do we face now?
9. What energy problems will future generations face?
10. How will the price of energy change in the future?
11. How should energy production and use be regulated?

10.6   Possible Projects

1. Prepare forecasts of energy supply and demand 5 and 20 years into the future. Support your projections with a discussion of the forces that will lead to changes. Consider both the USA and the world.
2. In September 2020, the Governor of California announced a proposal to ban sales of new gasoline-powered cars and trucks in the state after 2035. Describe the impacts that may arise as a result of the ban. Consider both the direct impacts due to reduced vehicle emissions and other impacts such as changes in the vehicle fleet, sources of energy for the replacement vehicles, and consequences for the electricity and petroleum sectors.

Sources

General: [1] BP Statistical Review of World Energy, June 2020

Figure

Number	Source
10.2	http://www.northeast.railfan.net/challenger.html

Index

© The Author(s), under exclusive license to Springer Nature Switzerland AG 2021
M. Cronshaw, *Energy in Perspective*,
https://doi.org/10.1007/978-3-030-63541-1

Printed in the United States
by Baker & Taylor Publisher Services